INSTALLING HEAT PUMPS

BY JACK JOHNSON &
WAYNE HAMMOCK

FIRST EDITION

FIRST PRINTING

Printed in the United States of America

Library of Congress Cataloging in Publication Data

Johnson, Jack, 1943-
Installing heat pumps.

Includes index.
1. Heat pumps. I. Hammock, Wayne, 1935-
II. Title.
TH7638.J63 1983 697 83-4864
ISBN 0-8306-0263-1
ISBN 0-8306-0163-5 (pbk.)

Contents

Introduction

Conservation is becoming extremely important because of the energy costs that we are being forced to pay. The choices that we have for heating our homes are many. Solar heat and wood stoves are alternatives.

Most types of energy-saving devices provide a level of saving. Consider whether the level of saving will be 1 percent or 90 percent of your total energy bill. Solar heat in many areas is a good investment, but how many cloudy days do you have in your area? Is the area too cold in the winter for solar heat? The weather service or your local power company can provide weather information.

Will your new energy system be able to pay itself off within seven years in reduced heating costs? Many fuel-saving devices on the market provide savings. When you consider payback on the investment, then questions arise. We feel that any energy-saving device must give a payback period of less than seven years.

A heat pump is part of the solar generation family. The sun in effect provides the heat for your home. The sun heats the air, and the heat in the air is removed by the heat pump. A conventional solar heating system uses a closed water or fluid circuit for conversion and transportation of the heat to the inside of the home. Solar heat is limited in the cooler areas of the country because of efficiency and its inability to air-condition during the summer.

Maintenance of a properly installed heat pump system will be very low in terms of time and parts cost. The heat pump's air filter must be checked often, or a premature compressor failure could result.

Heat pumps aren't new. They have been around for about 80 years and have been improving constantly. Solid-state electronics and refrigeration have become closely related in that more precise control is available. Because of the technical breakthrough in modern heat pump design, 50, 60 and 70 percent reductions are now being obtained by the average user.

The largest and best-known names in the heating and air conditioning industry are not making the needed steps toward better machines. The most efficient heat pumps have been developed by the smaller companies for the last three years. In selecting a heat pump it's very important to talk to as many people as possible, then buy the brand that has the best reputation. When you have chosen a pump, compare the efficiency figures for that unit to other similar-sized pumps. Compare one size to a similar size of another brand.

This book will assist you in partial or complete installation of a heat pump. By using it, you can save up to $800 on the cost of the installation.

If it wasn't for the understanding of our loved ones, this book wouldn't be in existence today. The long hours of planning and writing were difficult for our families, but they stood behind us giving encouragement and love. We dedicate this book to Carole and Darlene with love.

Chapter 1

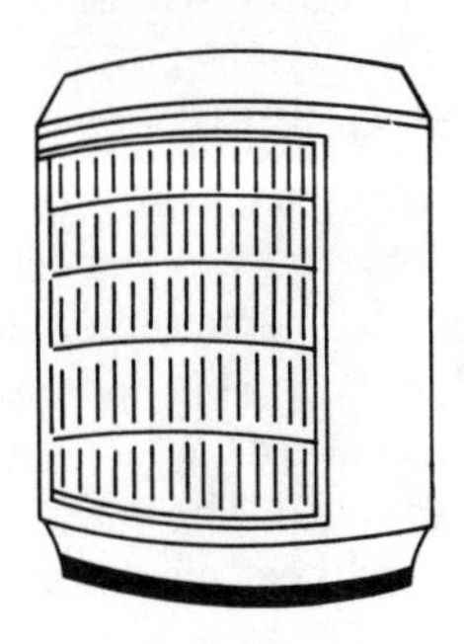

General Information

A heat pump pumps heat from one area to another. In the summer the pump extracts the heat from the indoor air and then expels that heat outside to provide cooling. When a heat pump is in the heating mode, the pump extracts heat from the outdoor air and transfers that heat to the inside of the house. The primary advantage of a heat pump is its ability to reduce heating costs. The reduction in cost averages 40 to 70 percent, depending on geographic areas and the average winter temperatures. The colder a given area, the less the saving factor.

The lower operating temperature limit of a heat pump varies between manufacturers because of design and materials. All heat pumps are not equal in their ability to save fuel dollars. The old law that says, "You pay for what you get," is more true than not. The ability of a pump to save money is rated. This rating is the coefficient of performance (COP) figure, and it will be covered to some extent later in this chapter.

Conservation and comfort are important benefits and deserve elaboration. A good pump system costs a little more than a conventional heating system with an air conditioner. The heat pump system costs so much less to operate that it actually pays for itself. The payback period can be as short as three years or as long as 10 years. In the Pacific Northwest five to seven years is average. You will obtain a return with a heat pump that is seldom found in the investment world. Another advantage of a heat pump is that it tends to even out temperatures between the rooms of your home. A cold room can be pure misery in the winter.

HEATING SYSTEMS AND EFFICIENCY

If you have an ideal furnace design that is operating at its best and is properly installed, you can expect the following. Oil and natural gas will produce 0.8 cents in heat for each penny spent on fuel. Electric heat will produce a one for one exchange. A heat pump will produce 2.3 and 3.1 cents of heat for each penny spent on fuel at 47° F., depending on the brand. The variation of 2.3 to 3.1 cents is a direct effect of design. All heat pumps are *not* equal; stay with a recommended brand.

The average heating systems found in the field are 60 to 70 percent efficient. Most of these average systems lose heat through the chimney because of a less than ideal fire. Besides the fire loss, we can add the loss of the heated space. Heating and building design are prime factors when we discuss heating costs. Costs are becoming so important that some design innovations are coming out of the laboratories. Instead of 80 percent efficiency, the new gas units are approaching the 100 percent level. A heat pump, or getting free fuel, is the only way to exceed a return of 100 percent for your fuel dollar. Fuel costs can be reduced by using solar energy or geothermal sources, but equipment cost and geographic area are the determining factors. Wind power is another source, but the equipment costs on an individual basis could be exorbitant.

A heat pump is a solar heating unit. The sun heats the air, and the pump extracts the heat from that air. Every object on the earth that is warmer than the air transfers heat to that air. A heat pump uses, in part, other people's wasted energy.

Heat pump costs vary. Use caution in selecting a unit. A few tens of dollars less in the initial purchase price may cost thousands of dollars in operating costs. In the selection of a quality pump, use the COP figure or the efficiency figure. No one will say his equipment is second or third best.

Heat pumps can be applied by themselves or with other types of heating systems such as gas, oil, or electric furnaces. The system will continually monitor the indoor temperature, giving just the right amount of heating or cooling for your comfort. If you are replacing equipment, building a new system, or just wanting to reduce your heating bill, a heat pump is the answer.

HEATING AND AIR CONDITIONING COMPANIES

How do you choose a good company to supply and service your pump? No one is going to tell you that he has equipment rated at number five on a scale of one to ten. No one is going to tell you about the built-in problems that occur because of equipment design. Even the best units will have small electrical and mechanical problems. Most dealers with second-rate equipment will try and sell you on price. The least expensive equipment is usually the worst.

All major manufacturers produce only the case and coils in the equipment. Most if not all of the parts inside the case have other companies' names on the parts. Don't attach much importance to any name brand. The largest name in the trade produces second-rate equipment in the opinion of many technicians.

Ask people you know about local heating firms. Check on the firms' reputations. The more questions you ask, the better off you will be. The heating dealers will only sell one or two name brands. They are not able to buy every brand at a competitive price. Take your time and shop around.

COEFFICIENT OF PERFORMANCE

When you decide that a heat pump would be a good investment, a means of comparison must be used for selection. The coefficient of performance or COP rating is a heat pump's efficiency rating. If a heat pump has a COP of 2.8, it means that for each penny invested to power the heat pump, a return of 2.8 cents in heat is delivered. The COP figure is also a function of outdoor temperature; the number changes with different outdoor temperatures. Forty-seven degrees Fahrenheit is a standard test temperature for the various manufacturers. If heat pump number one has a COP of 2.6 at 47° F., and heat pump number two has COP of 3.1 at 47° F., which would be the best choice? The second unit is by far superior and would give you ½ cent more in heat value than the first unit at 47° F. The COP figure is listed for every heat pump on the market. By using the efficiency figure at different listed outdoor temperatures, it's easy to choose your best buy. Even a tenth of a cent difference in the COP rating can mean a large saving when you look at the operating expense for 10 years.

The reason heat pumps are rated at different temperatures is because the output, or heat from the pump, is less at colder outdoor temperatures. The amount of heat in the air at 47° F. is greater than the amount of heat in the air at −10° F. The outdoor section of the heat pump moves a constant volume of air through the outdoor coil, and less heat is picked up from the atmosphere. If the outdoor coil was increased in size, more heat could be extracted. Technical problems can result in adding surface area to the coil.

The point that a heat pump return in saving diminishes below what it costs to operate the machine, plus the mechanical wear cost, is the cutoff point. Manufacturers suggest turning the heat pump off anywhere between 20° F., to −10° F. or even −20° F. The cutoff point tells you how efficient the machine is.

EER RATING

In the cooling mode the efficiency rating is stated as the EER or *energy efficiency ratio*. In both COP and EER ratings, the higher the ratio, the more efficient the heat pump or cooling system operates. Normally the COP is the most important, because that is the mode where the bulk of the savings will occur. If you are in the market for an air conditioner, compare the EER ratings of the various brands. If a dealer can not supply the EER rating, then you should find another dealer—one who knows his product.

APPEARANCE AND DESIGN

Most products in America are sold primarily on outward appearance rather than inner design and function. A round heat pump is not the best design. The round shape may be appealing, but that model isn't necessarily the best. The round heat pump is round so that people will buy it for the way that it looks.

A heat pump is a machine built by men for men, and it's not perfect. The efficiency levels of heat pumps are created by the engineers that design them, and the company executives give their approval. The designing process is a compromise among looks, material cost, and the effectiveness of the heat pump. If you want a good heat pump, compare the efficiency rating numbers or COP. A heat pump only has two primary functions—to keep you in your comfort zone and to save heating costs. The heat pump itself is of prime importance, but the installation of the pump can affect the overall performance. The efficiency rating for heat pumps is covered above in some detail under Coefficient of Performance.

METALS AND THEIR EFFECT ON HEAT PUMPS

Metals are used in heat pumps for mechanical strength and for their ability to conduct electricity and transfer heat. The exterior case of the heat pump should be steel to prevent denting in everyday use. The mechanical strength of the exterior case will also add years to the heat pump's effective service life. Steel does have one problem in that it rusts. If you buy a heat pump that has been painted using an electropainting process, the quality will be excellent. Ordinary spray painting is good, but it doesn't cover as well as the electro process. The quality of the paint will affect the life span of the unit. The heat pump can be waxed to extend its service life.

Besides the case on the outside unit of the heat pump, there is another exposed area. When the heat pump is in the heating or cooling mode, it draws air through the outdoor unit. The part of the unit that transfers the air temperature resembles the radiator of a car, and it is called a *coil*. The coil is constructed of hundreds of *fins* spaced far enough apart to allow air to travel between them.

The fins are made of aluminum. The fins on the radiator or coil are mechanically weak; they will bend readily. Protect the outside unit from abnormal abuse. If a cat or dog urinates on the fins, the fins will be eaten away because of a chemical action. The fins appear to have been eaten away by acid. A heat pump shouldn't be exposed to other unusual chemicals in the air. We've seen units almost vanish because of a chemical reaction that occurred with the metal. If in doubt, consult an expert in metals or possibly your fire department.

The fins in the coils are attached to tubing. The fins increase the air of the tubing so that the tubing is effectively exposed to more air temperature. By increasing the area of the tubing surface with the fins, the unit becomes more efficient. Copper conducts heat better than aluminum as proved in laboratory testing. If the fins also are made of copper and tubing, a maximum efficiency would be obtained over aluminum construction (Fig. 1-1).

Most coils used today are made of aluminum. A few companies are offering copper tubing with aluminum fins, and this feature is a money-saver. If you have a choice, choose a pump with the copper tubing because of the extra heat transfer advantage. This relates to the reduction of that operating expense.

ABSOLUTE ZERO

Absolute zero is a hypothetical temperature characterized by the complete absence of heat and is equivalent to approximately −273.15° C. or −459.67° F.

Today's heat pumps operate down to about −20° F. The lower operating limit differs between various manufacturers.

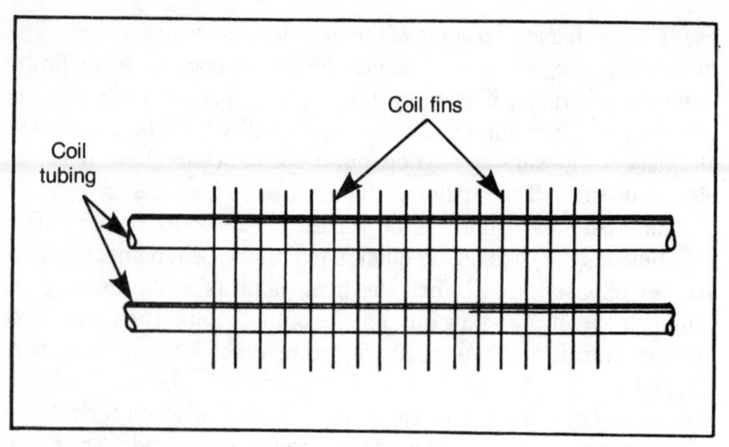

Fig. 1-1 The fins are attached to tubing.

ELECTRICAL SAFETY

All electrical power to the heat pump or related equipment must be turned *off* when working on the wiring. Under the right conditions, 0.016 ampere in electric current can kill. A 100-watt light uses enough power to kill 50 people if conditions such as moist skin exist. Unless you are qualified as an electrical installer, leave the power wiring to a qualified electrician.

REFRIGERATION SAFETY

Freon pressure inside the refrigeration circuit may reach very high levels during certain malfunctions. Pressures in excess of 500 pounds per square inch can be reached in little time. Do not heat or cut any refrigeration lines without the proper training. Do not attempt to charge or vary the charge of a heat pump without instruction. Never use automotive type cans of refrigerant to charge a heat pump. If a wrong connection is made, the cans can blow up. Leave the job to a professional.

Some refrigeration jobs require that the interconnecting copper lines be manufactured on the job site. When this is done, only silver solder is used. Soft lead solder will not hold the normal operational pressures. The lines will blow apart.

Refrigerant gas *(Freon)* is not toxic in small amounts in a natural state. When the gas passes through an open flame, it becomes very toxic and deadly. If a small amount of the toxic gas is inhaled, the effect is additive through several small doses. Various nervous disorders may result. Only use qualified people when the refrigerant system needs repair or adjustment, and they should never work in an unventilated room.

Several types of refrigerant gas are in use today. Most heat pumps use R-22, but sometimes R-500 or R-12 are present. Refrigerant types should not be mixed; the right kind has to be used for the particular machine. The type of refrigerant usually is noted on the nameplate of the equipment.

TYPICAL HEAT PUMP INSTALLATION

Figure 1-2 shows typical heat pump installation. The outdoor unit is on a snow or support pad. The refrigeration lines are fed through the wall to the area of the furnace. The turns in the refrigeration lines will be on a larger radius in the actual installation. The radius of the bends in precharged lines have to be made so as not to kink the copper lines. The larger the line with which you are working, the easier it will kink. Larger lines must be bent with a larger radius.

The indoor coil in Fig. 1-2 has been installed in the supply side of the furnace. Because the coil is in the supply, a special control system will be needed to allow either the furnace or the heat pump to operate—never both simultaneously. If the furnace and the heat pump function together, the temperature in the supply duct will be more than the heat pump can take. The heat pump will fail. When the coil is in the supply, a trap is not needed on the condensate drain.

BUILDER MODEL HEAT PUMPS AND AIR CONDITIONERS

When a contractor is building a home, the cost of the structure is very important. The more features a contractor can include in his building, the more appealing that building will be to a buyer. When the major manufacturers of air conditioning equipment produced the builders model, it was to open a new market. For the market to buy enough heat pumps to provide the profit that the manufacturers need, a less expensive model was devised for builders. The builder heat pumps that we have seen are difficult to work on, inexpensively built, and are generally substandard. Stay away from builder model heat pumps and air conditioners.

One of the best-known American air conditioning companies introduced a builder model a few years ago. If an electrical failure occurred in the right control circuit, the outside unit would fill with ice, then the outside case would blow up like a balloon from the

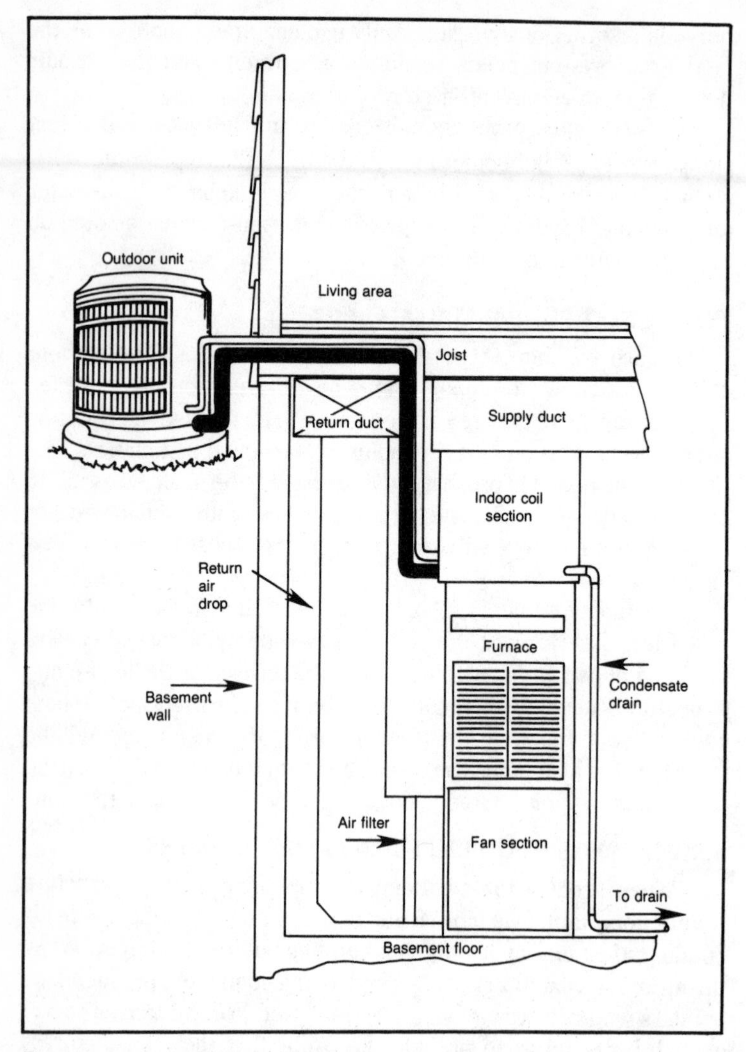

Fig. 1-2. A typical heat pump installation with interconnecting refrigeration lines between the outdoor unit and the indoor coil.

internal pressure of the ice. Eventually the unit will become extremely noisy and fall.

SIZING THE HEAT PUMP

The size of your heat pump is extremely important and will directly affect your comfort. It will also affect the ability of the unit to conserve fuel.

As in any profession, heating/cooling contractors vary in ability. The key to their success is proper schooling and vast experience in the field. Any engineering problem can be solved, but solving that problem effectively with minimal cost creates a give and take situation. You will need assistance in performing a heat loss and heat gain tabulation for your home. An estimator could be called and paid by the hour to perform this service. When the loss and gain are known, it's possible to size the heat pump.

A unit that is too large will add to the cost and give poor heating and cooling comfort. A unit that is too small just won't do the job. Remember that a heat pump is also an air conditioner, and it has to be the right size for your home or business.

The estimator will need to know the square footage of each floor, the size and direction that each window faces, the ceiling height on each floor, door size and location, and insulating values. Another important point is whether or not the building is airtight. What type of windows do you have in the home? If your home has a crawl space, does it have a vapor shield?

Heat flows from a warm area to a cooler area in any medium and always in the same direction. The rate that the heat flows from one area to another is governed by the temperature difference between the two areas. Insulation reduces the flow rate, according to its thickness and type. Water movement or electrical current flow is similar to heat flow or transfer.

Windows gain or lose heat rapidly. One square foot of single-pane glass will lose as much heat as 4 square feet of the average residence wall or 3.5 square feet for the average residence ceiling. The heat loss ratio of four to one is exhibited by a window in a worst case example. Double-pane windows reduce this lost factor. The addition of a storm window will reduce the heat loss in a single-pane window by 60 percent. Instead of using a single-pane window, and storm window, a double-pane window could be installed, which would be a 43 percent reduction in heat flow from that of a single-pane window.

BUILDING INSULATION

One inch of fiberglass equals 44 inches of concrete in insulation value. From these figures, you can see how concrete removes the heat from a dwelling.

Another item that uses large quantities of heat is a sliding glass door. The glass door is functional, and it even adds a nice effect to a room. It will use three times as much heat as a comparable window,

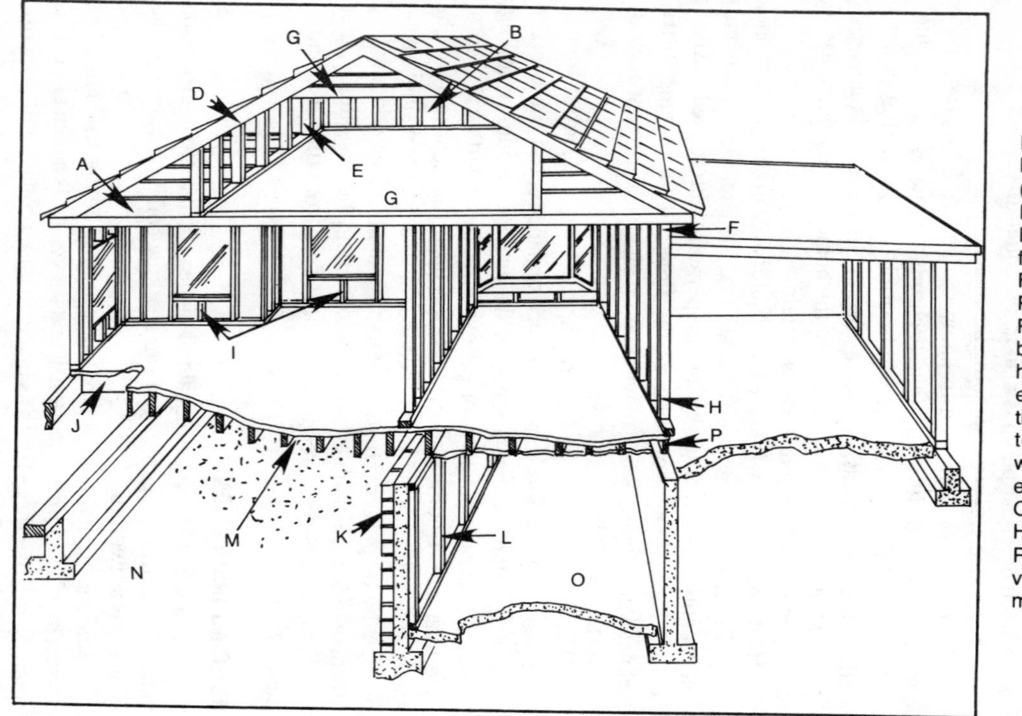

Fig. 1-3. Guidelines for insulating a home. (A) Ceiling joists—up to R-38. (B) Finished attic end walls—up to R-13. (C) Attic living area—up to R-13. (D) Rafters to knee wall in finished attic—up to R-19. (E) Finished attic knee walls—up to R-19. (F) Short exterior walls—up to R-13. (G) Finished attic collar beams—up to R-38. (H) Wall to unheated garage—up to R-13. (I) All exterior walls—up to R-13. (J) Cantilever area—up to R-19. (K) Sill—up to R-19. (L) Heated basement walls—up to R-11. (M) Under-floor exposed to cold—up to R-19. (N) Open crawl space—up to R-19. (O) Heated basement—up to R-19. (P) Rim joist—up to R-19. Note that R values are the maximums recommended for cost effectiveness.

though. The main reason a glass door causes a greater heat loss is the seals between the sections. The more the door is used, the greater the loss.

Metal window and door frames conduct heat at even a greater rate than concrete. Wooden frames are by far the best choice.

Caulking around windows and door frames is important. The heat in your home can literally flow outside through cracks in the building. We recommend a yearly inspection of your home.

WHERE TO INSULATE

When you insulate a home, the effect usually is to separate the warm areas from the colder areas outside and within the home (Fig. 1-3). The addition of insulation to any area of the home should be cost effective in that it returns the cost of the insulation in heating and air conditioning costs in a reasonable amount of time. Achieving recommended R values will ensure the saving of money and energy, with added comfort. R values are a measure of a material's ability to slow heat transfer. The higher the R value, the better the insulation. Different materials may offer widely varying R values. When buying insulation, consider whether or not the material is flammable.

Chapter 2

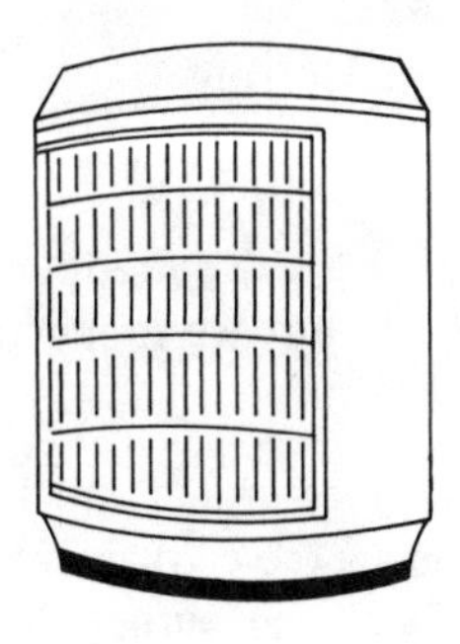

Furnace Air Flow Systems

The air flow through the furnace or air handler of a heat pump system is somewhat critical. Depending on the manufacturer's specifications, the air required will be from 280 to 450 cubic feet of air per minute through the indoor coil. The wide variance is due to the construction of the indoor coil.

The air flow is measured as shown later in the formula, or the temperature of the liquid line can be felt for an approximation of air volume. When the outdoor temperature is about 40° F., the liquid line should be warm to the touch. If the line feels hot, the air flow has to be increased by changing the size of the motor pulley or by increasing the horsepower rating of the motor if it is a direct drive fan assembly. The amperage draw of the motor should be checked when fan speed is increased to prevent a motor overload. If the fan speed is increased without a noticeable effect on the liquid line temperature, then a duct problem exists in the system. When duct problems arise, the solution usually involves finding a restricted section of return air duct and increasing the size of the smallest opening. Check the point that the return air duct enters the wall return grille. Compare square-inch opening sizes.

FORMULA FOR DETERMINING AIR FLOW IN A FORCED AIR FURNACE

$$\text{Air flow in cubic feet per minute (CFM)} = \frac{\text{output Btus (British thermal units) of the furnace}}{\text{output duct temperature minus return duct temperature}}$$

Example:

$$\text{CFM} = \frac{80{,}000 \text{ Btu (furnace output)}}{150° \text{ F.} - 70° \text{ F.}}$$

$$\text{CFM} = \frac{80{,}000 \text{ Btu}}{80° \text{ F. (difference)}}$$

CFM = 1,000 cubic feet per minute of air.
The furnace is moving this quantity of air.

The input duct temperature is the temperature of the air returned from the rooms via the cold air return duct to the furnace and through the air filter. The output duct temperature is the temperature of the warm air supplied to the various rooms in the home. Use only a quality thermometer in making this measurement. When measuring the output temperature, get just far enough downstream in the supply duct so the thermometer can not "see" the inside of the furnace. A false reading will occur if the thermometer can "see" the furnace because of infrared radiation.

Figures 2-1 through 2-3 are examples of air flow restrictions in the air duct and ways of correcting them. Arrows indicate air flow. Figures 2-4 through 2-6 show air movement in a commonly used air duct.

SYSTEM DETERMINATION

There are five different types of furnaces, air handlers, and air distribution systems: *upflow, downflow, horizontal, lowboy,* and *heat pump coil with an air handler*.

When installed in a basement the upflow uses the floor of the living area for the supply register and return grille location. The upflow can be installed on the floor of the main living area if there is no basement. The ceiling is used for the register and grille locations. The air duct is located in the attic space. When installing your heat pump coil in either case, it would be installed as described in Chapter 3.

The downflow model is usually installed on the floor of the main living area, generally with a crawl space underneath the floor. Possibly your home was built on a cement slab; in either case the coil would be installed in the same manner. The top of a downflow furnace is the return air side. If you install the coil in the return, it would set on top using the furnace as the coil base. If you live in a high humidity area, there is an alternate installation method with

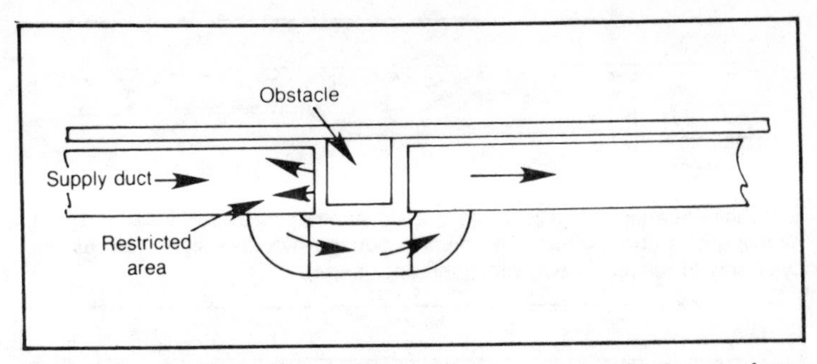

Fig. 2-1. The diagram illustrates how obstacles are overcome by the use of poor patchwork duct design. Notice how the air flow bounces off the flat plate before passing around the obstacle. When air flow is opposed by turbulence, friction is increased, causing reduced overall volume within the ducting.

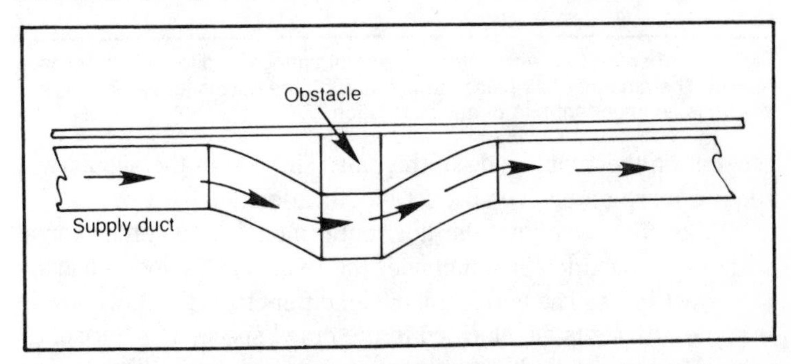

Fig. 2-2. Smooth directional changes, when properly sized, reduce friction loss and assure a maximum amount of air flow through the duct.

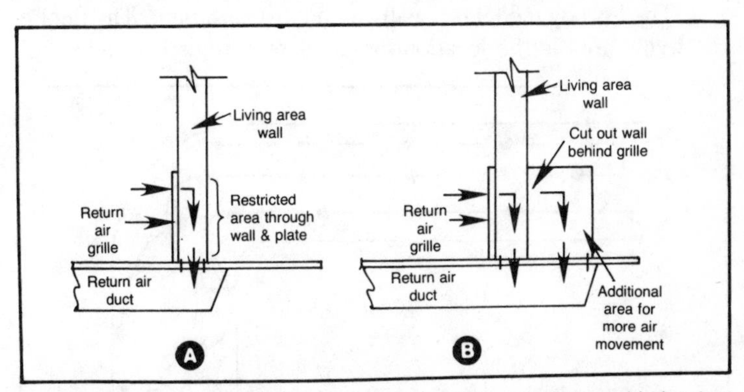

Fig. 2-3. (A) When a residential heating system is installed, the cold air return grille(s) is installed on an inside wall. The interior of the wall acts as part of the return air duct, connecting the air flow through the cutout wall plate in the wall and through the floor. Severe air flow restriction occurs in this type of installation because of the small sized opening through the floor. (B) Increasing the area of the return opening through the floor has solved many air flow problems.

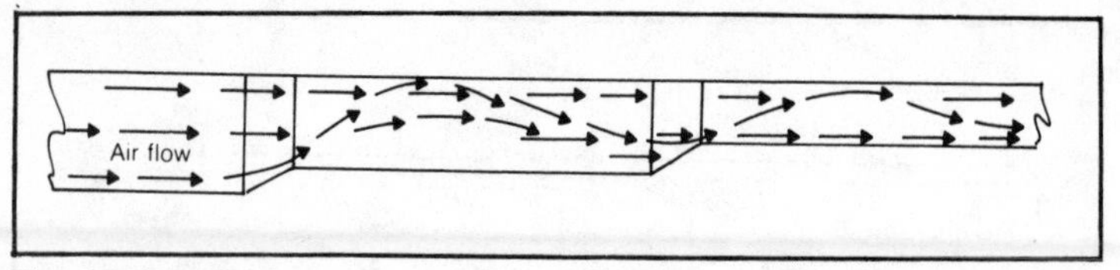

Fig. 2-4. Tapering one size of air duct to another size "streamlines" the air handling of a duct system. The primary concern when designing an air duct system is to reduce friction within the enclosure.

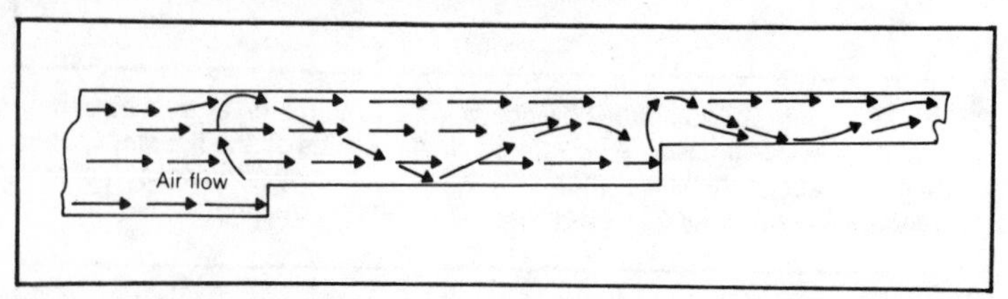

Fig. 2-5. When air flow encounters an abrupt change of direction, turbulence results. The amount of air being carried through the duct is lessened. Air duct design is an important part of any installation.

the coil on the supply side of the unit. The coil in the supply will require more extensive air duct modifications.

Like the downflow, the horizontal model is generally found with the distribution system under the living area's floor—usually in a crawl space. The horizontal model differs from the downflow in that the furnace is also located in the crawl space. The horizontal may also be installed in an attic space using the ceiling as the location for the supply registers and return grilles.

The lowboy model is usually found in a basement. The floor of the living area is the location for the supply registers and return

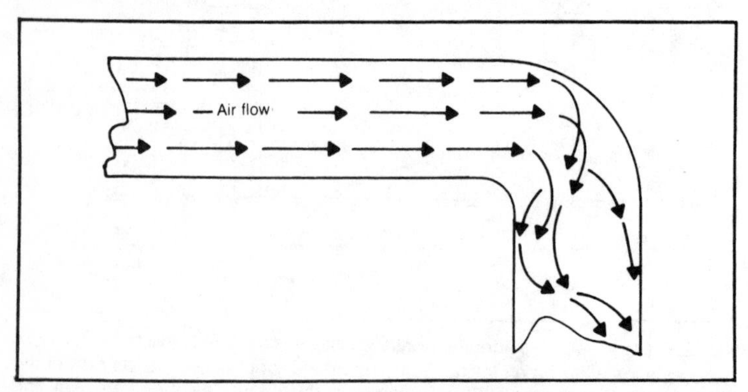

Fig. 2-6. It wouldn't seem that a 90° change of duct direction would cause a problem. This is not the case in an air system. Direction changes should be kept to an absolute minimum.

grilles. Many of the older model furnaces were of the lowboy design—very well-built and durable. It's relatively simple to adapt the heat pump coil to this model because of the ready-made base on the furnace.

The heat pump coil with an air handler is the type of installation you would use if you have baseboard electric or hot water heat in your home. These systems do not require an air duct to distribute the heat, so you would need a complete air duct distribution system installed for your heat pump. This book does not cover the installation of a complete air duct system, but you can obtain the information required by consulting your local heating specialist.

PHYSICAL CHARACTERISTICS OF HEAT PUMP A AND SLAB COILS

The A coil is not as versatile as the slab coil, but it is much more efficient. It has approximately 20 percent more surface area of coil, depending on manufacturing design. This coil can be used on upflow or downflow heating systems (Fig. 2-7).

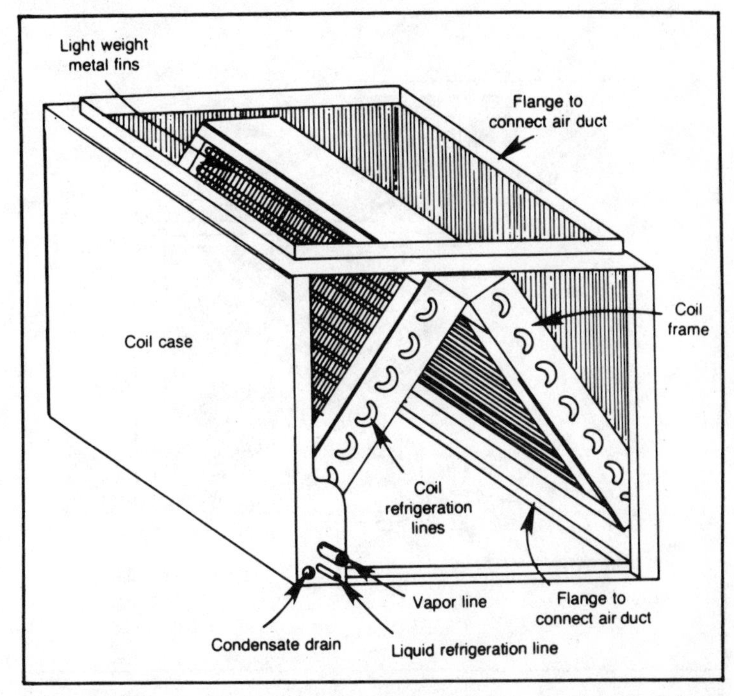

Fig. 2-7. A vertical A coil that is used in heat pump and air conditioning systems. The condensate drain outlet will be at the lowest point on the casing when installed.

The slab coil is more versatile than an A coil. Notice the optional condensate drain outlets (Fig. 2-8). The coil can be used for either horizontal air flow or vertical airflow. The coil turned on its side would be horizontal, so you would use the horizontal condensate drain. If it is upright, you would use the vertical drain. Install a plug in the drain hole that's not used. The slab coil is not as efficient as an A coil in that it has less coil surface.

COIL LOCATION

The relative humidity level in your area will dictate whether the indoor coil is installed in the supply or the return air duct. If you have an electric furnace, the coil will always be installed in the return air duct. Condensation on the heat exchanging surface of an electric furnace can't cause a rust problem, leading to premature failure. The elements of an electric furnace are constructed of materials that resist rusting. The entire element is in the cool, air

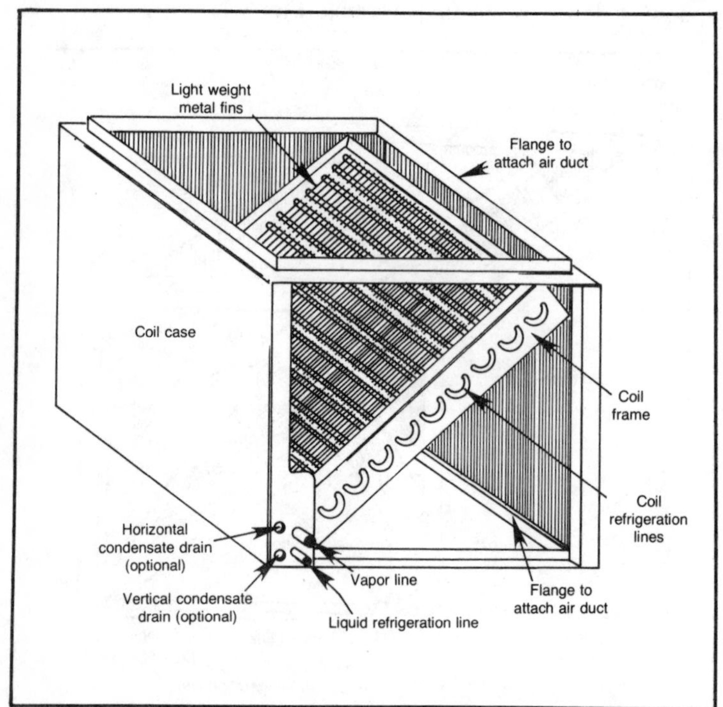

Fig. 2-8. Slab coils are supplied with many air conditioning and heat pump units. This type of coil isn't as efficient as the A coil.

conditioned air. The element doesn't have one side of its surface exposed to warm humid air, so condensation doesn't form on it. Condensation can be a costly problem if thought isn't given to a particular installation. Gas and oil furnaces may have the coil installed in the supply duct, depending on the geographic area. If you live in an arid or semiarid climate, placing the indoor coil in the return air duct is acceptable and recommended. Otherwise, install the indoor coil in the supply duct only.

TYPICAL UPFLOW FURNACE INSTALLATION

The upflow furnace draws the return air from the living room area down through the return air drop, then through the air filter and into the fan case of the furnace (Fig. 2-9). The fan then pushes the air across the heat exchanger of the furnace and into the supply ducts where it is distributed to the registers and circulated to the living area of the home. It repeats the same process until the thermostat is satisfied.

TYPICAL DOWNFLOW FURNACE INSTALLATION

The downflow furnace is generally installed on the floor of the main living area (Fig. 2-10). It draws the return air from the ceiling into the return air ductwork, then through the filter and into the fan case of the furnace. The fan pushes the air across the heat exchanger where it is heated and pushed into the supply air duct and distributed through the house. The process is repeated until the thermostat is satisfied.

TYPICAL HORIZONTAL FURNACE INSTALLATION

The horizontal furnace is generally installed under the floor of the dwelling (Fig. 2-11). It could also be installed in an attic area, with the supply and return registers located in the ceiling.

TYPICAL LOWBOY FURNACE INSTALLATION

The lowboy furnace is identical to the upflow unit, except that the fan and heat exchanger are in a horizontal position (Fig. 2-12). The unit pulls return air from the floor of the living area into the return air ductwork, then into the return plenum through an air filter and into the fan section. The fan pushes the air into the heat exchanger section where it's heated and sent back to the supply air

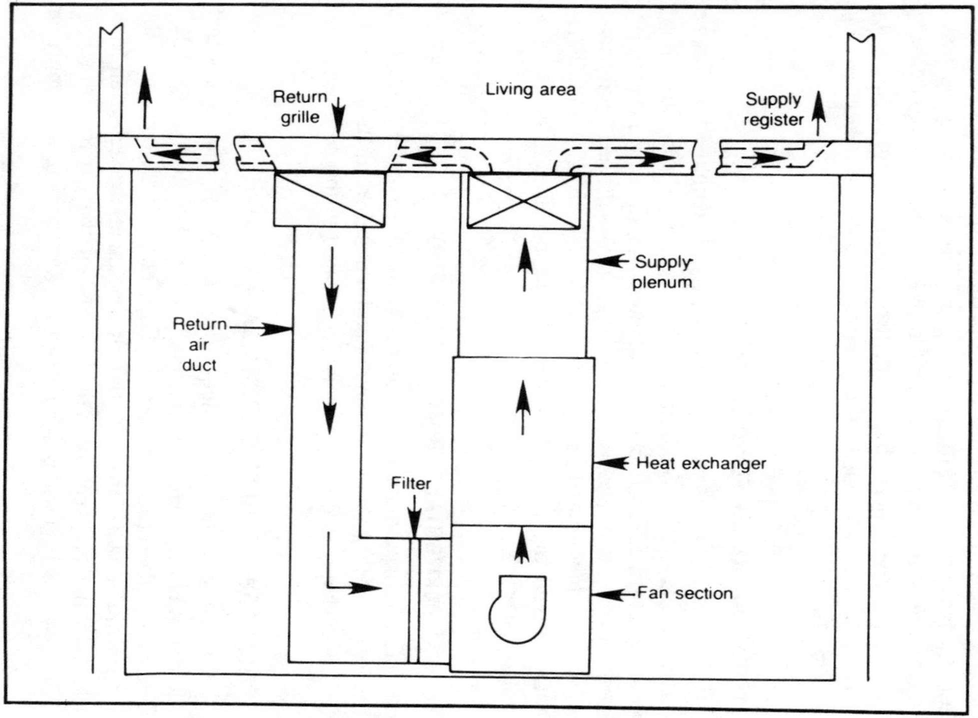

Fig. 2-9. An upflow furnace system directs the air flow through the furnace in an upward direction.

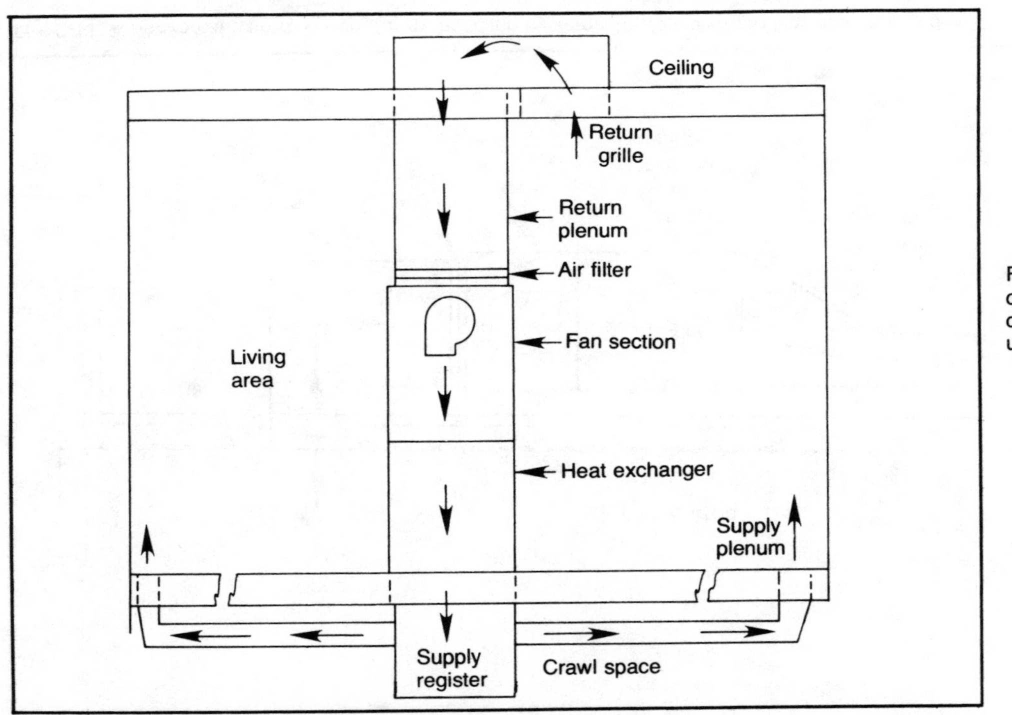

Fig. 2-10. Air enters through the top of the downflow furnace and then is discharged from the bottom of the unit.

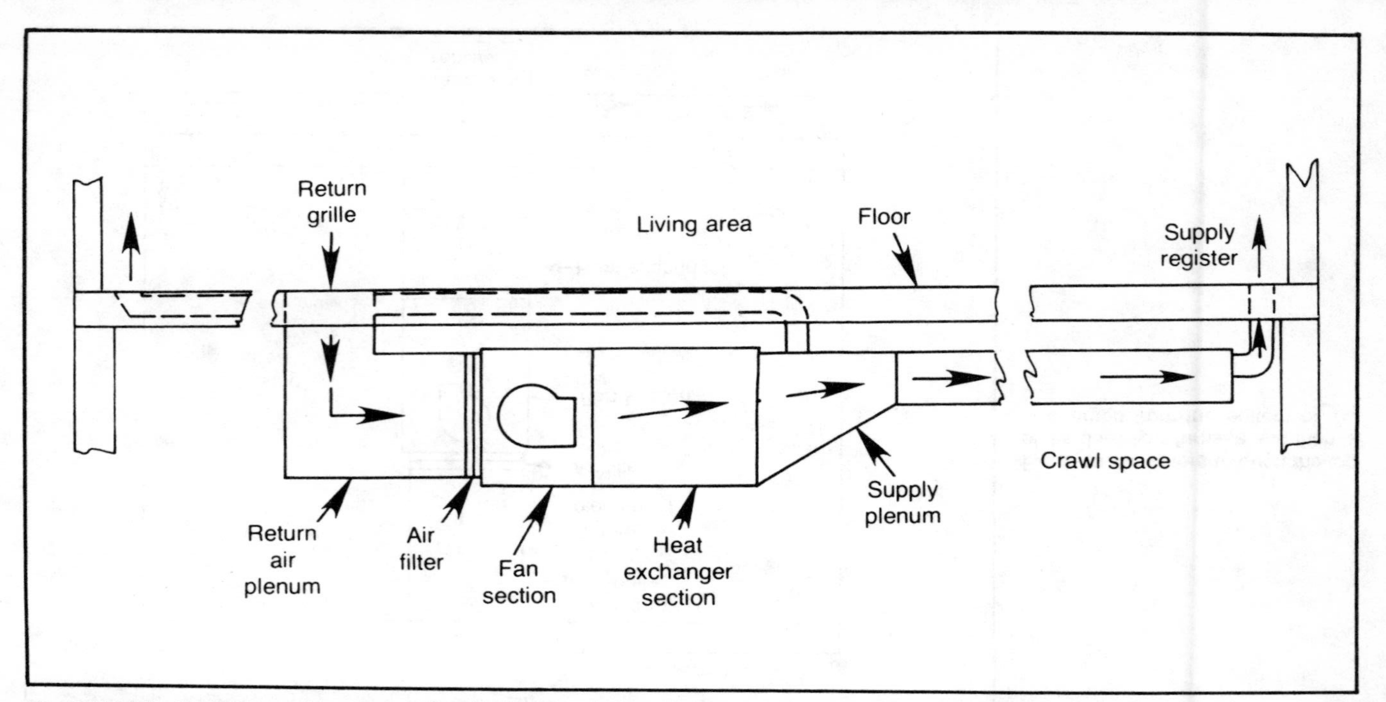

Fig. 2-11. A horizontal type of heating or air handling system is installed where limited space is available for the equipment.

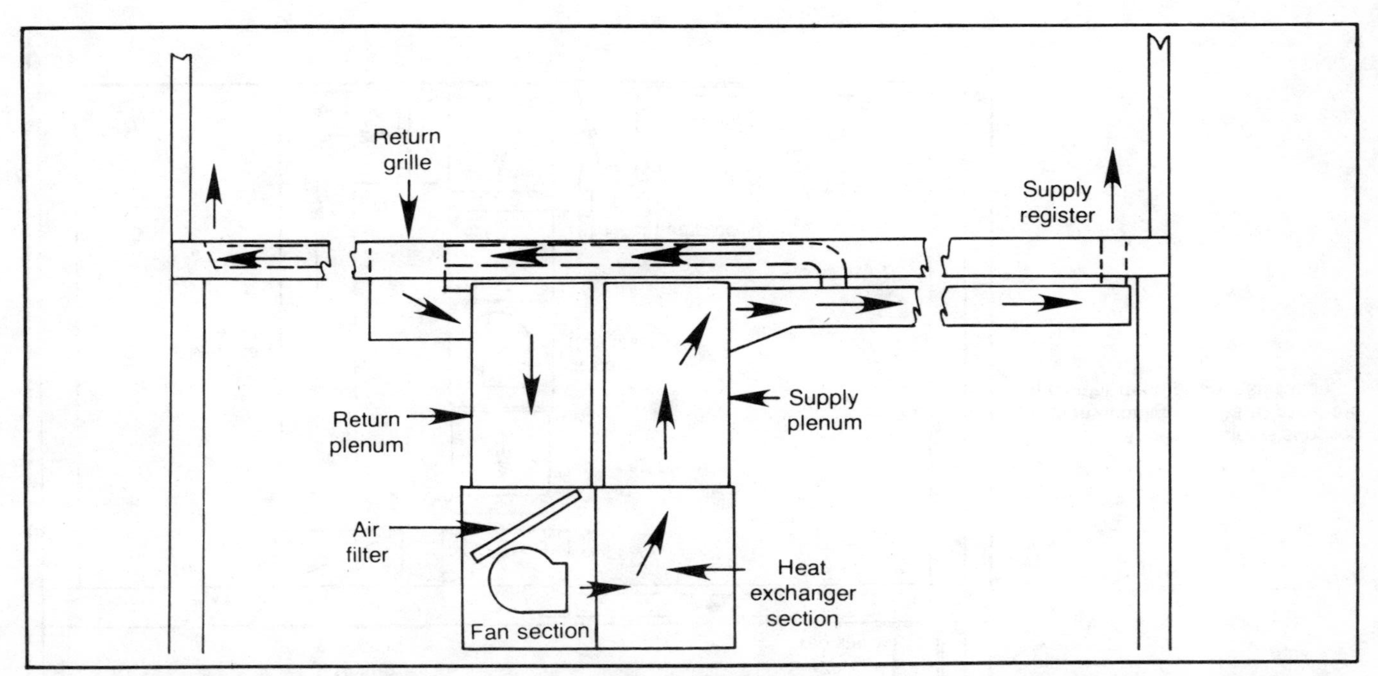

Fig. 2-12. A lowboy configuration is common in many older installations. The basic cost of a lowboy unit has made it less popular with today's contractors.

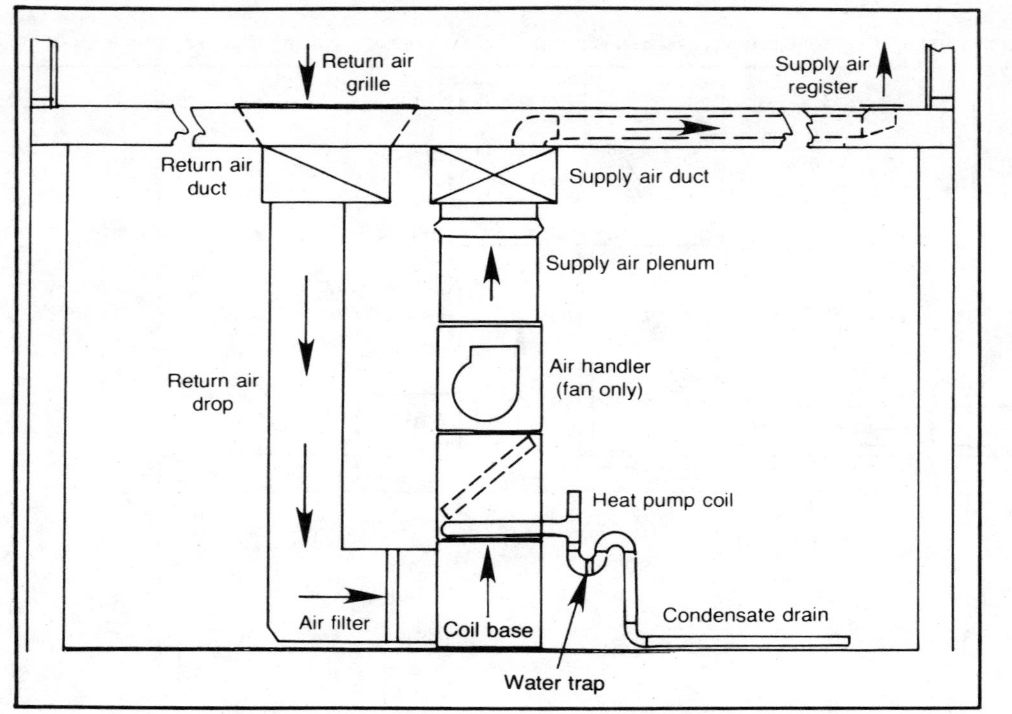

Fig. 2-13. An air handler is supplied in many configurations to meet the particular need of an installation.

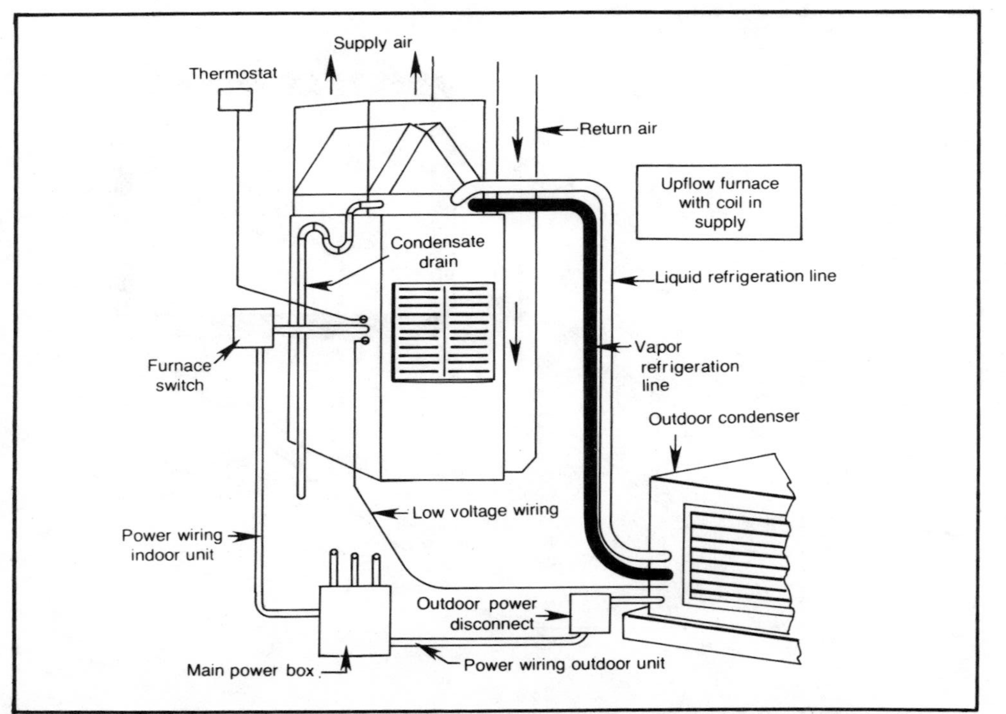

Fig. 2-14. Split system heat pump and air conditioning systems are interconnected electrically and also through the refrigeration piping.

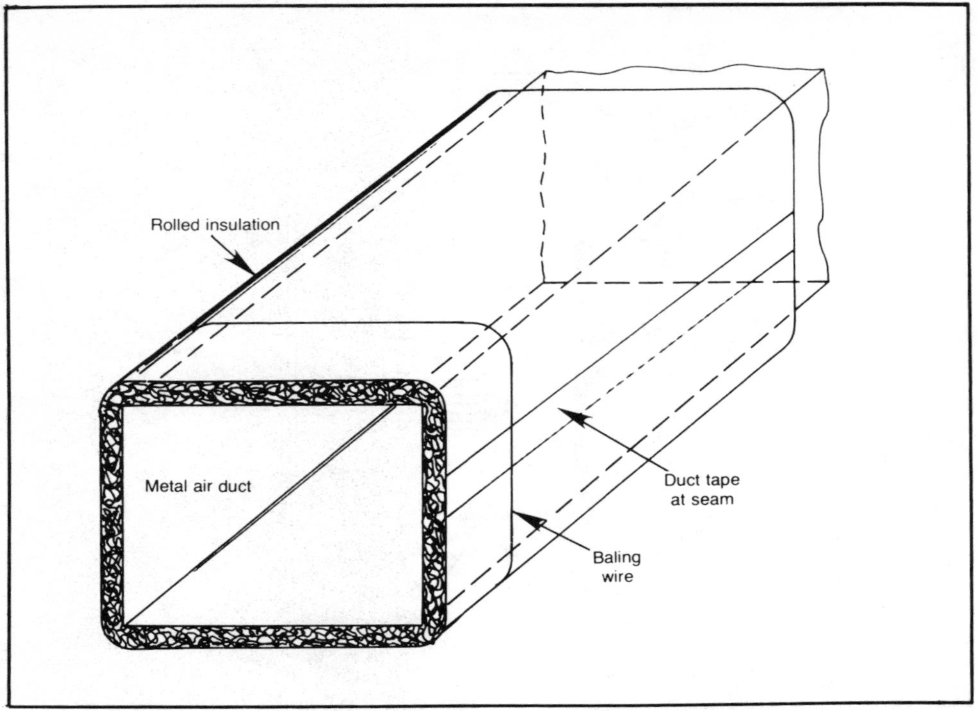

Fig. 2-15. Insulated ducting is a definite plus when considering energy savings.

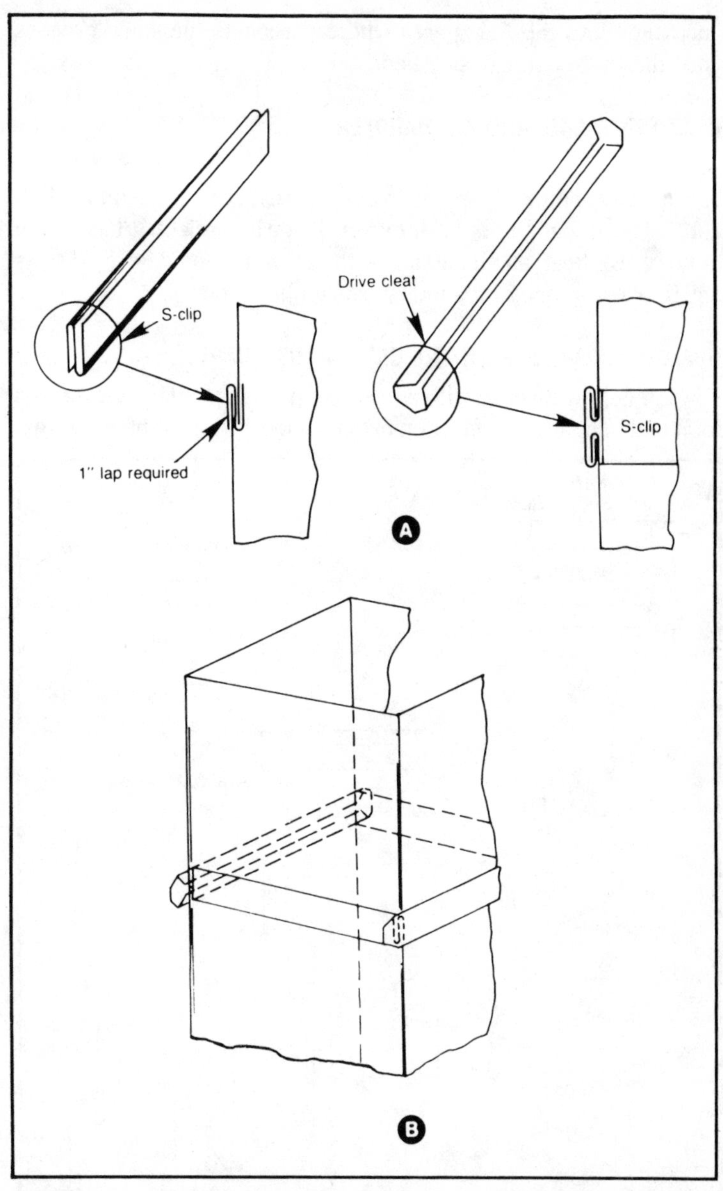

Fig. 2-16. (A) S-clips and drive cleats are used to connect sections of duct together. Note that the S-clip holds the sections apart. The drive cleat holds the sections of duct material together. (B) An isometric view of an air duct with the S-clips and drive cleats installed. Notice that two sides of the duct use drive cleats. The other two sides use S-clips. The ends of the drive cleats are bent around the corner to close the open gap found at each corner.

duct, then into the living area where it repeats the same process until the thermostat is satisfied.

HEAT PUMP COIL AND AIR HANDLER

An air handler is a case that resembles the casing of a furnace, but the air handler doesn't have any internal heating element (Fig. 2-13). The air handler is an air mover. If you have electric baseboard or hot water heat, an air handler with ductwork would be used so as not to waste money on another backup heat source.

UPFLOW FURNACE WITH THE COIL IN THE SUPPLY

An upflow furnace with a heat pump is shown in Fig. 2-14 as a total unit. The interconnection of the various parts is shown so that

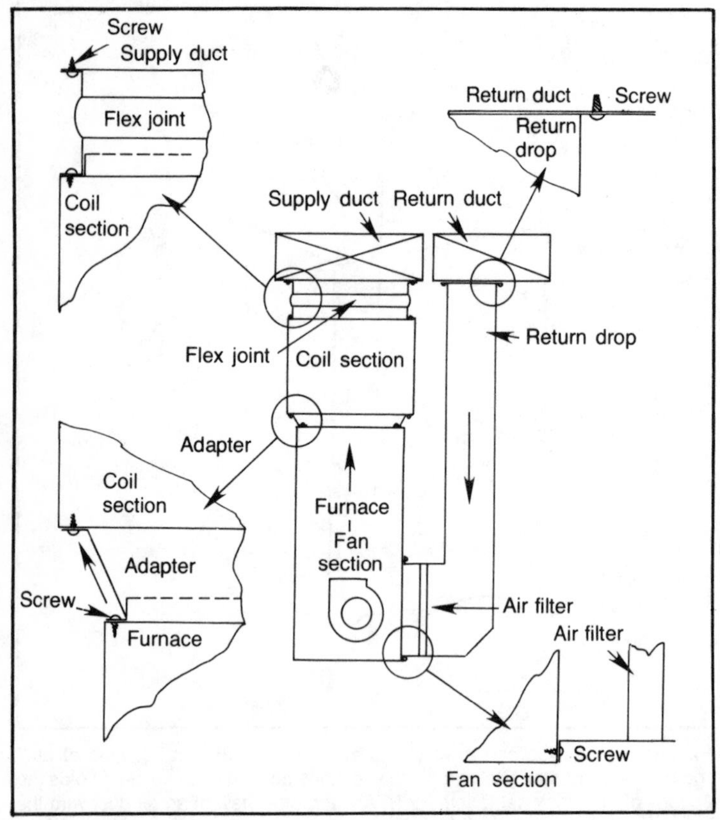

Fig. 2-17. Different types of duct require interconnection to form a solid unit. Silicone seal can be used to reduce air leakage and noise in joints near the furnace or air handler.

you can see what the system should look like after completion. The heat pump will require 230 volts. A gas or oil furnace will use only 115 volts for power. A trap is shown on the condensate drain of the coil. When the coil is installed in the supply side of the furnace, a trap is not necessary because the air is being pushed across the coil, unlike a return air installation that can suck the condensation out of the drainpipe and create a backflow problem.

WRAPPING INSULATION AROUND AN AIR DUCT

An air duct that is exposed to an unconditioned area such as a crawl space or attic should be wrapped with insulation to eliminate the radiation of heating or cooling, depending on the season in which you are using the heat pump (Fig. 2-15). The insulation also eliminates condensation that will occur due to the temperature differential. Ways of connecting duct are shown in Figs. 2-16 and 2-17.

Chapter 3

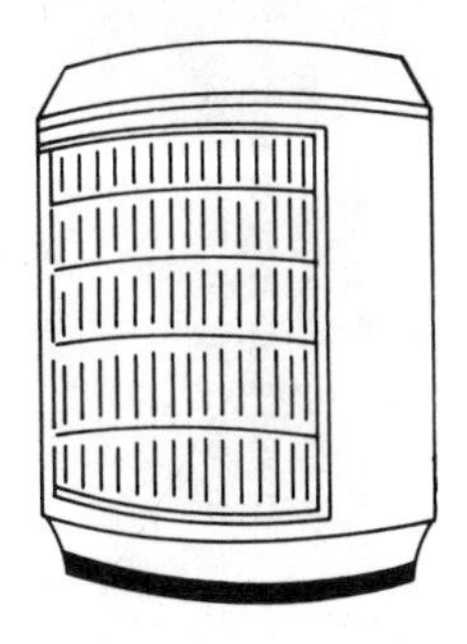

Duct System Installation

Installing the coil to your system is not as complicated as it may seem. Figure 3-1 shows a typical upflow system before a heat pump coil is installed. Figure 3-2 illustrates an upflow system with the necessary fittings to adapt a coil to the unit. Refer to these drawings often while you are determining the fittings needed for your installation.

ADAPTING A COIL TO AN UPFLOW SYSTEM (COIL IN THE RETURN)

The heat pump coil for this installation will take the place of the return air drop. Measure the distance from the bottom of the return air duct to the floor. This is the total distance you will have to install the coil and the sheet metal fittings.

The required fittings—including the coil base, coil case, filter rack, return adapter or transition, and flex connector—are shown in Fig. 3-2. The flex connector makes it easier to connect the final joint and reduces any vibration noise that might occur from fan operation.

Let's start with a hypothetical dimension of 86″ from the bottom of the return duct to the floor. You now know the space that you have for the installation of the five pieces needed. The dimensions on your particular system may vary.

There are two other measurements to consider before deciding the physical dimensions of the base. First, we must know the width and height of the fan section of the furnace. Second, we must know the width and depth of the outlet side of the coil case. Let's

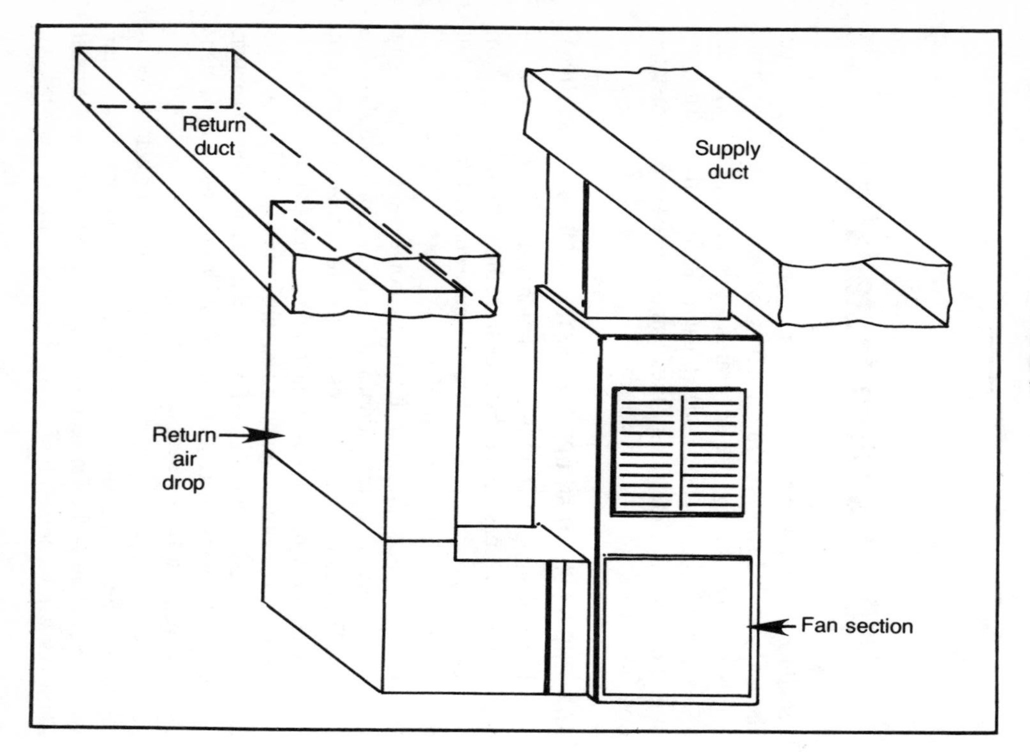

Fig. 3-1. Furnace duct systems are nothing more than a series of hollow boxes and tubes connected together to form an air passage.

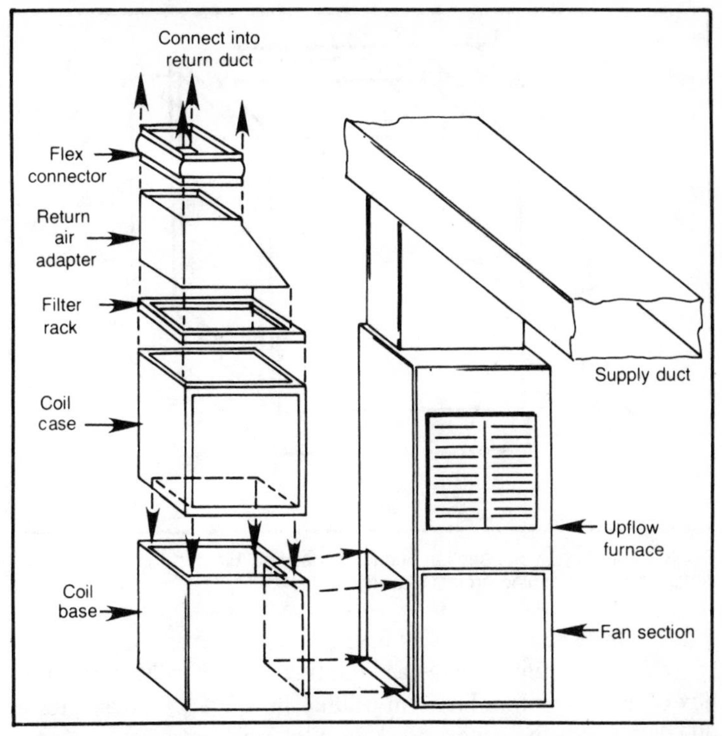

Fig. 3-2. Adding a heat pump coil to a furnace duct system requires using several adapters to interconnect the components. Be sure that all fittings are large enough to carry the required CFM of air needed by your heat pump.

assume the width of the fan section is 24″, the height is 20″, and the coil case outlet dimension is 22″ × 22″. Figure 3-3 shows the proper dimensions for your base.

Your furnace will more than likely have a smaller opening cut through the side of the fan case. You will have to cut it to the size of the coil base opening. There will be a 1″ flange on all four sides that is screwed to the fan case section. There is a 1″ flange on the four sides of the top on which the coil case sits.

Assuming the height of the coil case is 26″, we have now used 48″ of the 86″ with which we had to start. A 22″ height of the base plus a 26″ height of the coil case equals 48″.

Next we will consider the filter rack dimensions. Stock filters come in certain sizes, and you should use a stock size. Check this with your local hardware store or heating dealer.

Let's assume the inlet side of the coil case is 20″ wide × 20″ deep. Figure 3-4 shows the proper dimensions for the filter rack.

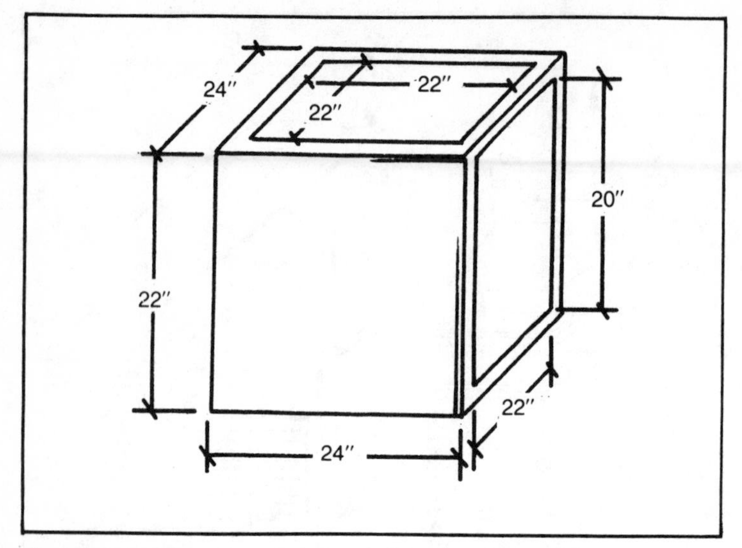

Fig. 3-3. A coil case or a section of duct has several dimensions that you must consider when planning an installation.

We have now used 50″ of the 86″, which leaves us with 36″ of space for the return adapter and the flex connector. Manufactured flex connector material is 6″ in width. The length of the adapter is 30″ plus 1″ for the connection. See Fig. 3-5.

The size of the return air drop on your system will dictate the size of the flex connector circumference. Let's assume that the drop is 12″ wide and 20″ deep. Figure 3-6 illustrates the sizes of the flex connector and the return adapter.

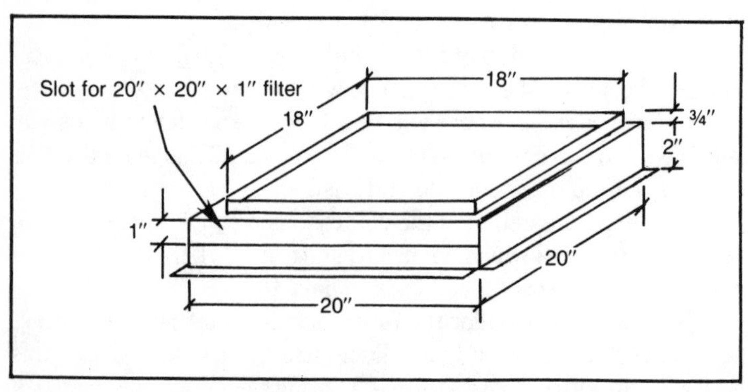

Fig. 3-4. All air entering a coil must be filtered. Usually workers at a sheet metal shop must build a filter rack to fit your installation. If you have the space, use a 2″ thick filter rather than a 1″ thick unit.

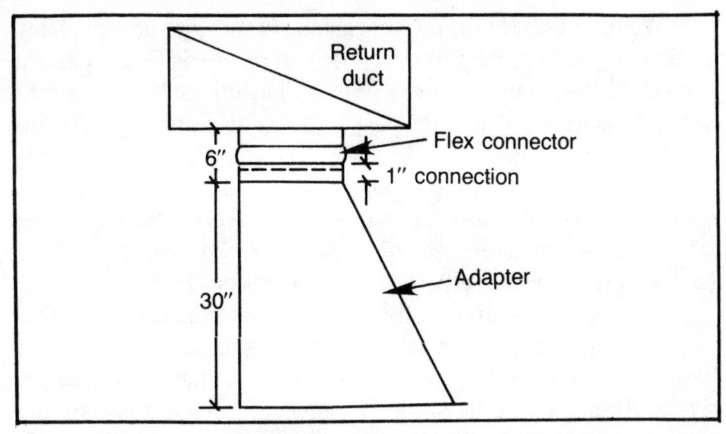

Fig. 3-5. Adapters are required when changing duct size for any reason. Your sheet metal shop will know the correct shape for any adapter that you might need.

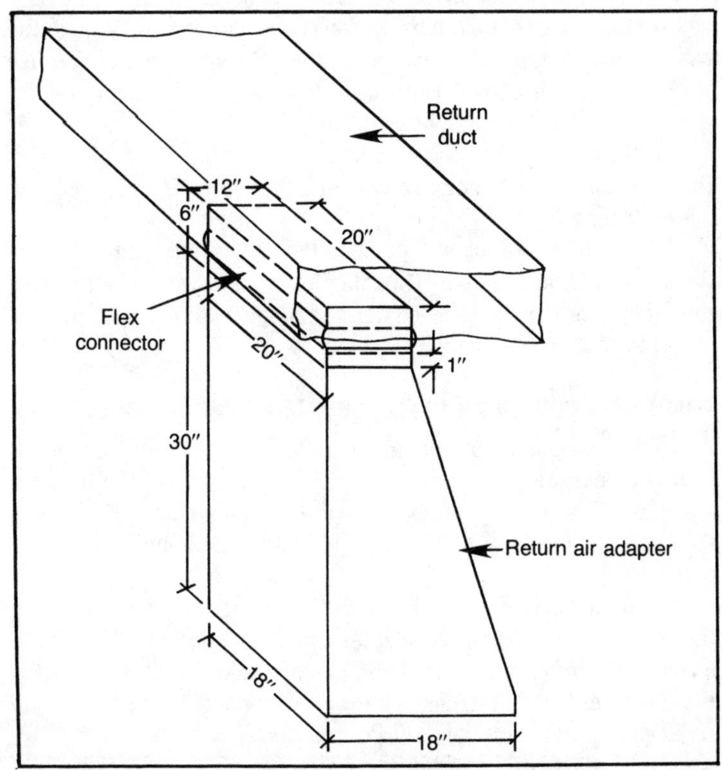

Fig. 3-6. Flex. connectors isolate the furnace from the duct system mechanically (to reduce noise). Small differences of alignment can be handled within the connector.

We have used fictitious dimensions in this instructional information. By substituting your particular measurements, you can compose a complete illustrated fitting list, then take it to a local sheet metal shop and have the people there quote a price to fabricate the fittings for you.

When you pick up the fittings at the shop, have the mechanic explain how to make the seam connection between adapter and flex. He will more than likely recommend using S-clips and drives. If so, have him turn the drive pockets, on the fittings for you.

The first fitting to install is the coil base. Cut the hole in the side of the fan section to match the opening in the side of the coil base. You can screw the base to the side of the furnace or use pop rivets. After the base is securely fastened, run a bead of silicone caulk around the seam. Air leakage between fan and coil must be kept to a minimum.

Run a bead of silicon caulk around the flange, on top of the base and set the coil case on the base. You can secure the coil case to the base with a couple of screws after the silicone dries. If you use screws, be sure not to drill into any refrigeration lines to the coil. Be careful while drilling any holes in the coil case.

After the coil is in place, secure the filter bracket to the top of the case using sheet metal screws or pop rivets. Again, be careful while drilling holes.

The return air loop will be installed on top of the filter rack by the same method as above. Then the flex connector is installed. Use the S-clips and drives between flex and adapter and between the flex and return duct.

ADAPTING A COIL TO AN UPFLOW SYSTEM (COIL IN THE SUPPLY)

The heat pump coil for this installation takes the place of the supply plenum. Measure the distance from the top of the furnace, which is the supply, to the bottom of the supply ductwork. This is the total distance you will have to install the coil and sheet metal parts.

Refer again to Fig. 3-2. It shows the required fittings, including the coil case, coil to furnace adapter, and flex connector. The coil to furnace adapter is called a transition in the trade. The flex connector is not a necessity, but it makes it easier to connect the final joint and reduces the vibration noise of fan rotation.

Let's start with an arbitrary dimension of 36″ from the top of the furnace to the bottom of the supply duct. You now know the space you will have for the installation of the sheet metal fittings and the coil case. The dimensions of your particular system may vary.

We will start with the coil to furnace adapter measurement. Let's first consider the total of 36″ with which we have to work. We will use a height of 26″ for the coil base. This is not necessarily the size of your coil case, but we will use it as an example. Manufactured flex connector material is 6″ wide. We have two unchangeable measurements—26″ for the height of the coil case and 6″ for the flex connector. We also want a ½″ flange bent to a 90° angle on the bottom of the flex, so 26″ plus 6″ minus ½″ equals 31½″. Also 31½″ subtracted from 36″ equals 4½″. The height of the coil to the furnace adapter will be 4½″.

Let's assume that the supply outlet on the furnace is 19″ wide and 16″ deep, and the inlet or bottom of the coil case opening is 20″ wide and 20″ deep. You need a 19″ × 16″ to a 20″ × 20″ transitional coil to the furnace adapter. See Fig. 3-7.

Notice that the adapter is 19″ wide on the bottom to fit the width of the furnace outlet. It is 22″ wide at the top to fit the outside measurement of the coil case, and there is a 20″ measurement inside the flange to fit the inlet opening of the coil case. The adapter is 16″ deep on the bottom, 22″ deep on top, and 20″ deep on the inside of the flange. The left diagram in Fig. 3-8 depicts the front view of the adapter, and the right diagram shows the side view.

Please take note that the increased depth of the transitional adapter is placed to the rear. This is so the fitting will not interfere with the vent on an oil or gas furnace. If you are adapting to an electric furnace, this offset will be immaterial.

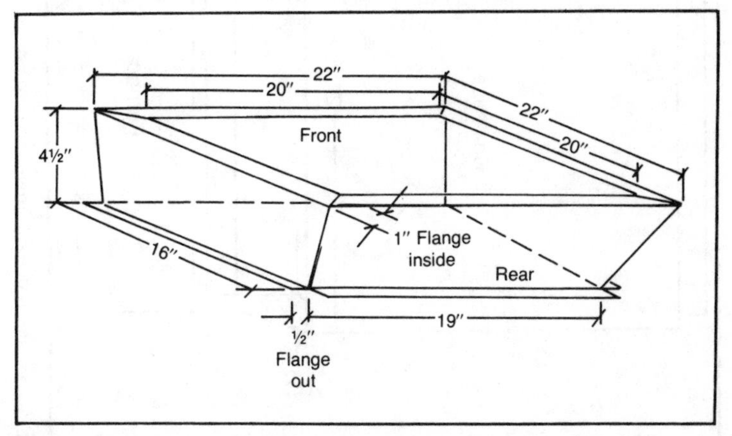

Fig. 3-7. An accurate diagram will be required to build a particular fitting for your duct system. Never reduce the size of a coil case opening, as this will cause an air flow restriction.

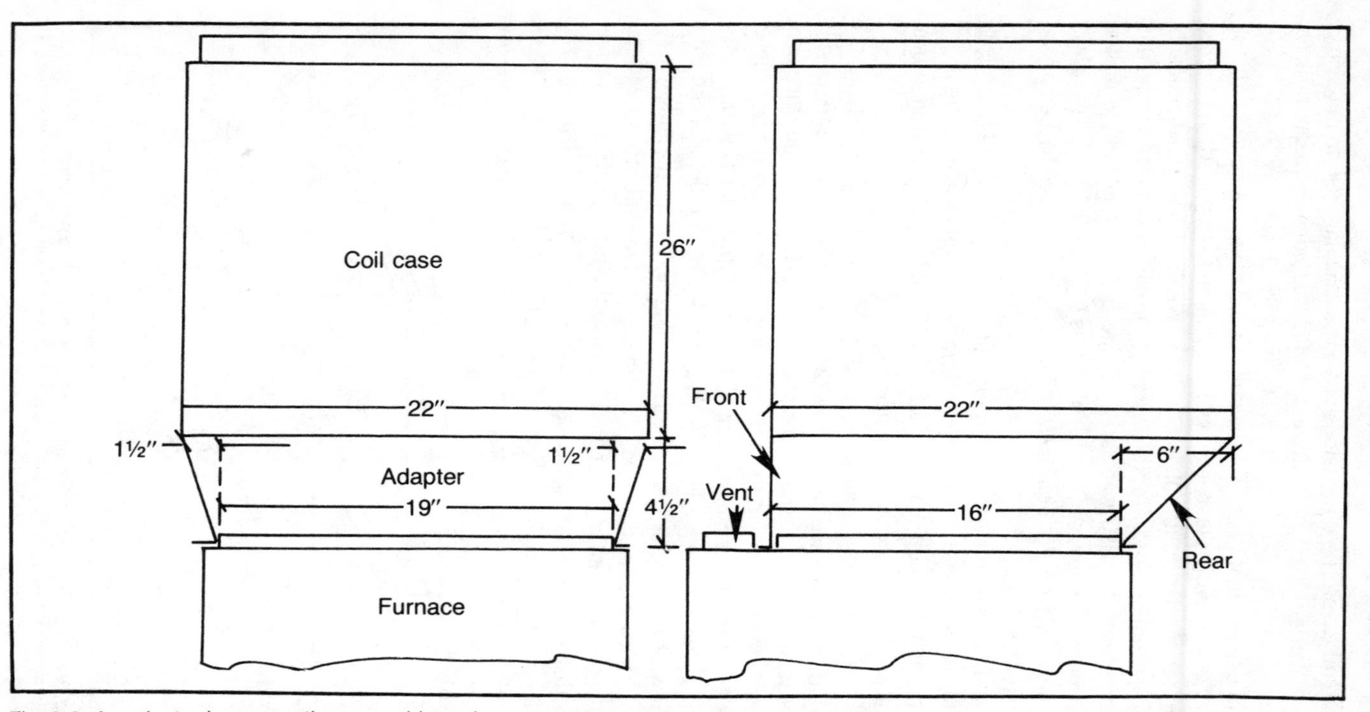

Fig. 3-8. An adapter has more than one side and may require a detailed drawing to explain its shape.

Now let's decide what dimensions we need for the flex connector. The stock width of the flex is 6″. We want a ½″ flange bent to a 90° angle at the bottom, so we will net out with a height of 5½″. Let's also assume the outlet side of the coil case is the same as the inlet; on some models these measurements do vary. We have established the size needed for the flex connector—20″ × 20″ × 5½″ high. See Fig. 3-9.

When you pick up the fittings at the shop, be sure to have the mechanic explain how to make the seam connection between the flex connector and the supply duct. He will probably recommend using S-clips and drives. If so, have him bend the drive pockets on the fittings for you.

The first fitting that we will install is the coil to furnace adapter. Set the adapter on the top of the furnace with the increased portion to the rear if it is an oil or gas furnace. Drill sheet metal screw holes through the fitting and furnace. Use a ⅛″ bit and insert #7 × ½″ screws. Make sure the fitting is tight to the furnace housing for a good seal.

Run a bead of silicone caulking around the circumference of the flange on top of the adapter. Before it sets up, position your coil case on the flange. By removing the access door on the coil case, you can reach inside and drill holes for more sheet metal screws. Be very careful not to drill into the coil or refrigeration lines. Three screws on each side are sufficient to hold the unit.

When the coil case is secured, you are ready to prepare the supply duct for the flex fitting connection. By using the information

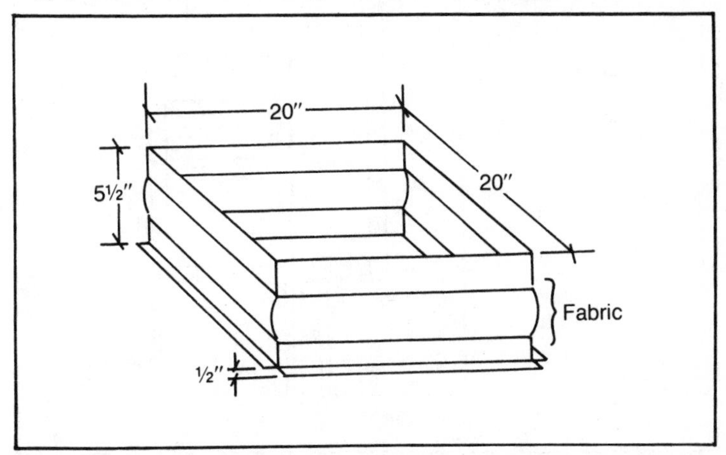

Fig. 3-9. A fabric flex connector has two sizes when the height is considered. One size is when the fabric is fully extended, and the other is when it is collapsed.

you received from the shop mechanic, there shouldn't be any trouble making a good fit. After this is done, insert more screws through the fitting and into the flange of the coil case.

Most upflow installations will resemble that in Fig. 3-10. This installation is without a heat pump coil. Figure 3-11 shows the component sheet metal fittings that will be required to adapt the heat pump coil to the supply side of an upflow furnace. Be sure that all fittings are large enough to carry the required CFM of air that your heat pump needs.

ADAPTING A COIL AND COUNTERFLOW SYSTEM (COIL IN THE RETURN DROP BELOW THE CEILING)

Figure 3-2 shows a typical counterflow installation before the coil is installed. Figure 3-13 illustrates the coil and counterflow furnace after the heat pump coil is installed. Refer to Figs. 3-12 and

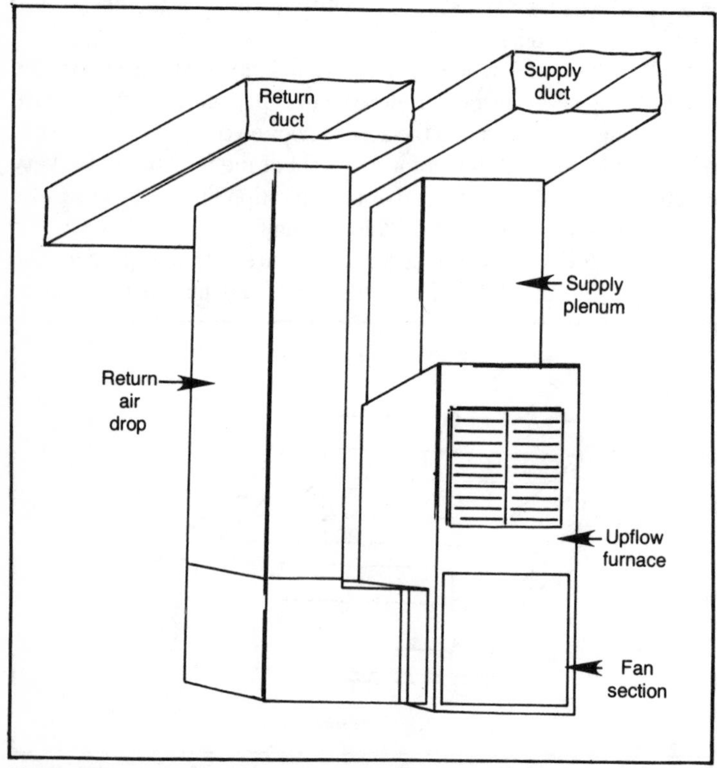

Fig. 3-10. The major components are usually the same, but the measurements will be different from one furnace installation to the next.

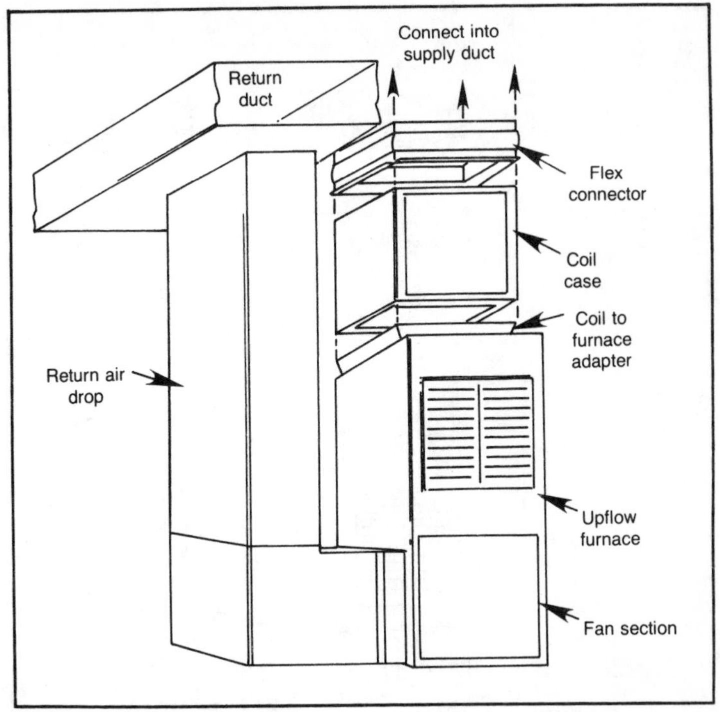

Fig. 3-11. Adding a coil to an upflow furnace will require several adapters, plus a special control system to operate the furnace or heat pump—never both at the same time.

3-13 often while determining the fittings needed for your installation. Notice in Fig. 3-13 that the coil is installed between the ceiling and the return inlet of the furnace. This is very handy for servicing, but if the space between ceiling and furnace is limited, you may choose the optional method of installing the coil in the attic space. The optional method is detailed completely later in this chapter.

Let's first determine the measurement of the return drop, which is the distance from the ceiling to the top of the furnace. Let's say it is 45″, and the return plenum where it penetrates the ceiling is 20″ wide and 18″ deep. The furnace return inlet flange is also 20″ wide and 18″ deep. The furnace return inlet flange is also 20″ wide and 18″ deep. The heat pump coil case is 24″ high. The inlet flange is 20″ wide and 20″ deep on the coil case. The outlet of the coil case is also 20″ wide and 20″ deep.

The three unchangeable measurements are the height of the coil case, the amount of return air duct that has to be left below the

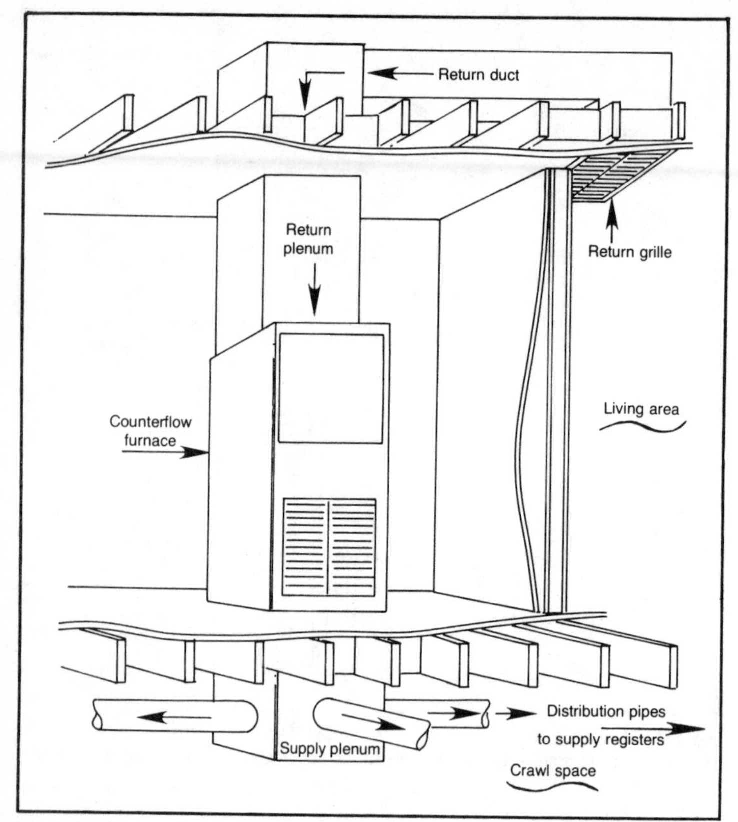

Fig. 3-12. Counterflow furnaces may supply air through an unconditioned space. Insulate the duct if needed.

ceiling for us to connect the fitting, also the height of the filter rack, which must be a minimum of 3″. Using these dimensions above, we can draw a rough sketch of the fittings needed. See Fig. 3-14.

Now that we know the height of each fitting, we can draw a rough shop sketch depicting each particular fitting. Let's start with the adapter between coil and furnace. Figure 3-15 will show you exactly what the fitting will look like. Figure 3-16 shows the filter rack.

The flex connector attaches to the top of the filter rack with S-clips and drives. Figure 3-17 is a flex connector fitting detail diagram.

The adapter between the flex and the return drop cutoff point will also be attached with S-clips and drives to both the flex and the return. Figure 3-18 details the adapter.

By substituting your particular dimensions, you can detail your own installation fitting requirements. Draw up a rough sketch following the directions we have explained. You can make a fitting drawing that you can take to your local sheet metal shop and then have the fittings fabricated.

When you pick up the fittings at the shop, have the mechanic explain how the S-clip and drive connections are made. Have him turn the drive pockets on the fittings for you.

When installing the fittings, refer to Fig. 3-13. The first fitting to install will be the coil to furnace adapter. Turn the coil case upside down and lay a bead of silicone caulking around the flange of

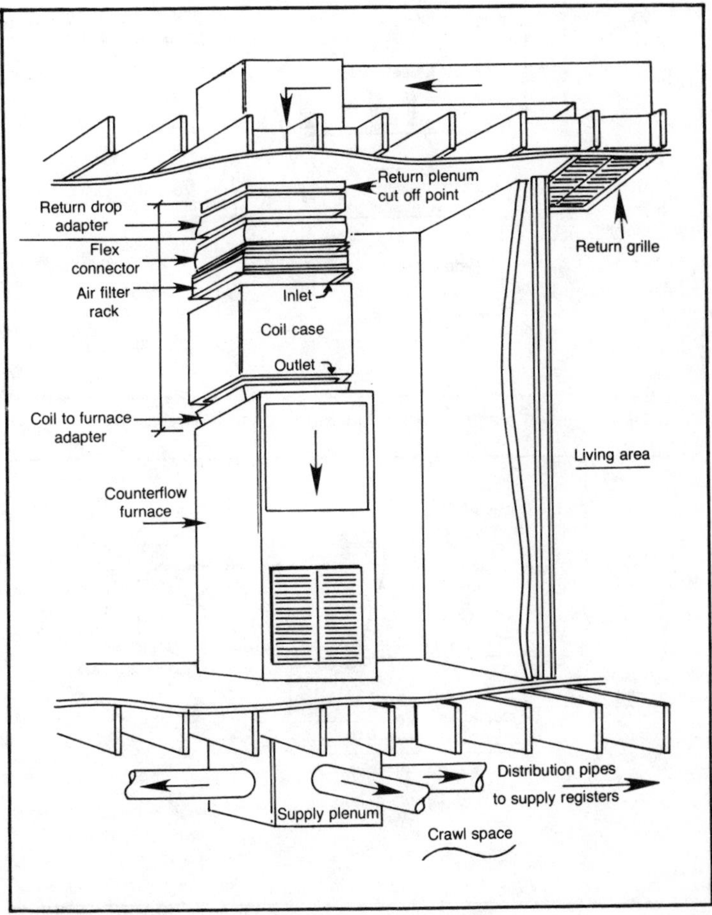

Fig. 3-13. Adding a coil to a counterflow furnace can be difficult at best when space in the ductwork is limited.

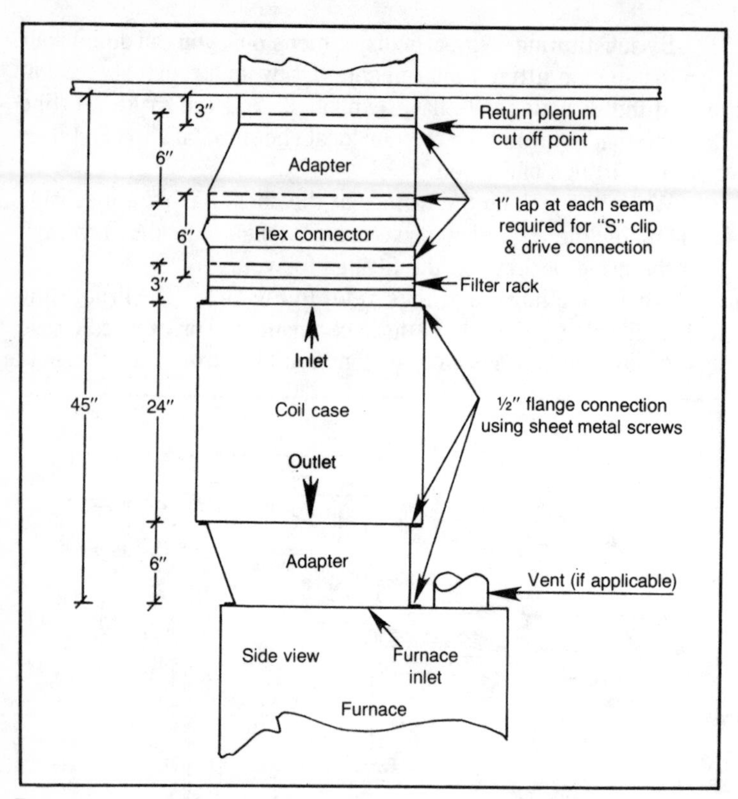

Fig. 3-14. A detailed plan is needed to ensure a good installation.

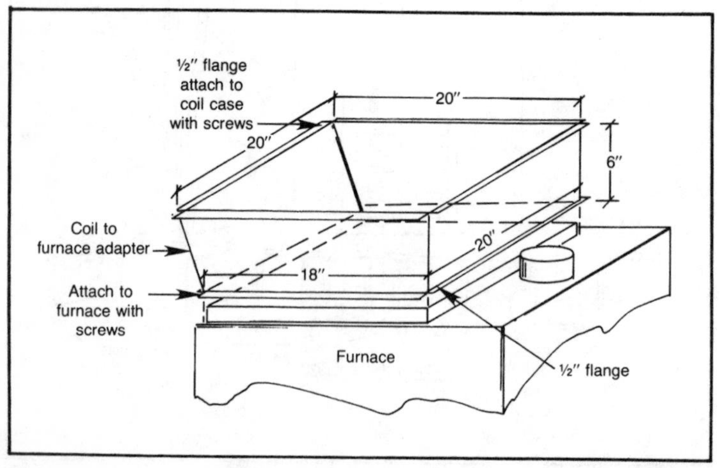

Fig. 3-15. Fittings are installed at the furnace or air handler first. Additional fittings are added, working away from the furnace.

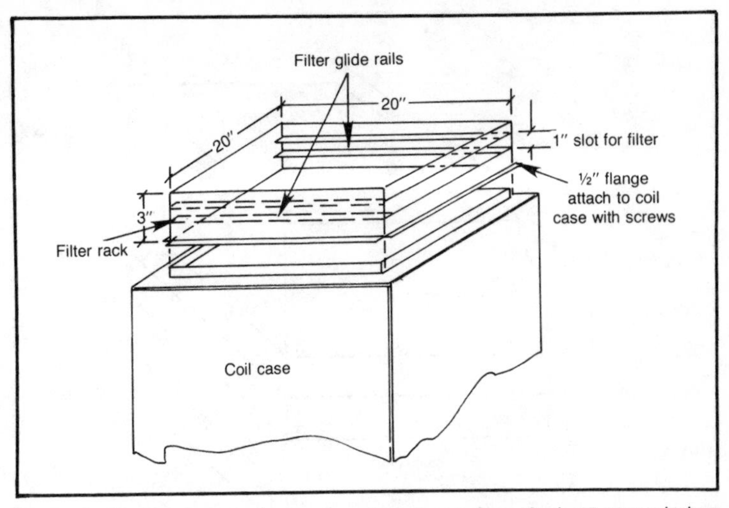

Fig. 3-16. Filter racks are always placed upstream from the heat pump indoor coil. Dust must be cleaned from the air before that air goes through the coil. If you have room for a 2″ thick filter, the cleaning efficiency will be increased.

the outlet side of the coil case. Turn the adapter so that the straight side is lined up with the front of the coil case. We want the expanded portion of the adapter to the back so as not to interfere with the vent (if you have gas or oil). Set the adapter on the bead of silicone and secure it to the case with #7 × ½″ sheet metal screws. Be very

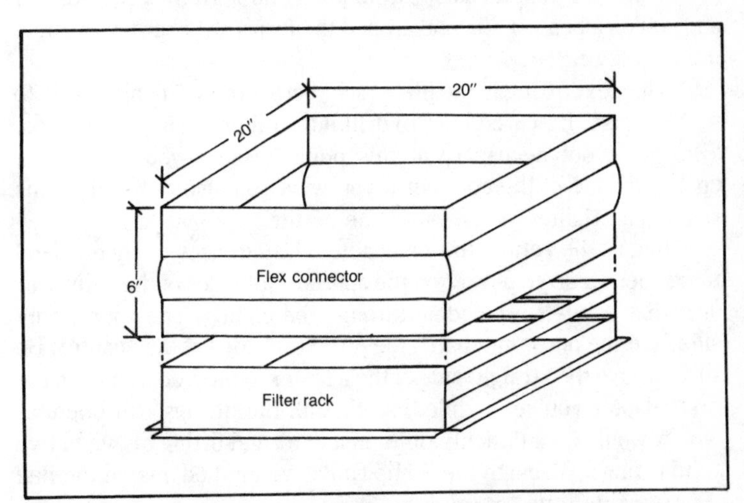

Fig. 3-17. A flex connector is usually applied on the main duct side of a duct system. The coil and the adapters are connected to the furnace, then comes the flex connector to isolate the furnace/coil assembly.

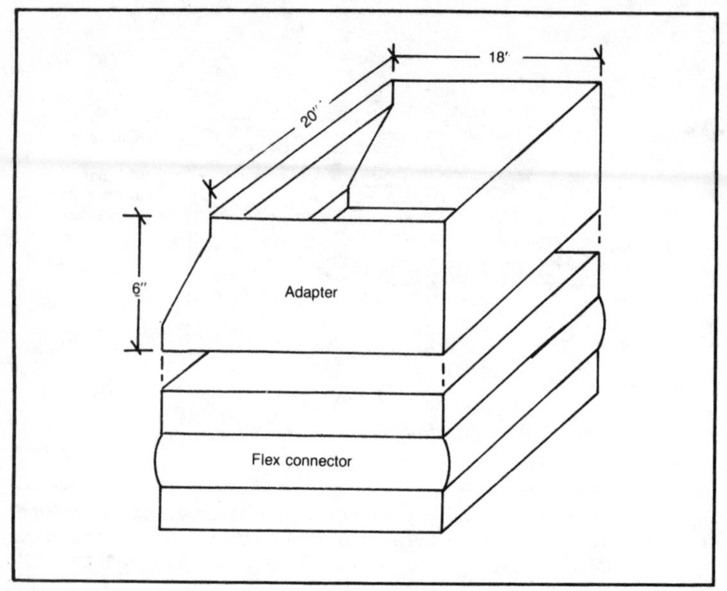

Fig. 3-18. Noise reduction is attained by the fabric in the flex connector so long as it isolates the vibration to the main furnace assembly.

careful not to drill holes into any part of the coil or refrigeration lines.

When the adapter is secured, place the bottom of the adapter and coil on a bead of silicone around the furnace. Secure with sheet metal screws.

The next fitting is the filter rack. Screw the ½″ flanged side to the coil case. Be careful not to drill into coil or refrigeration lines. Silicone is not necessary at this point because you are on the upstream side of the coil, but if you wish you also can seal at this seam. The tighter your seams, the better.

Install the return to flex adapter. Using a set of sheet metal tongs, bend a drive pocket on the appropriate sides of the return air duct that you left protruding through the ceiling. The appropriate sides are the ones that match the pocketed sides of the adapter. Be sure to have the straight side of the adapter to the front to match the first adapter you've installed, so that all the fittings will line up.

It would be a difficult job to install a rigid fitting between two solid fittings. By using the S-clip and drive method, install the flex connector into the return drop.

We recommend that all air duct in an unconditioned area such as attic and crawl space be covered with insulation.

Figure 3-13 illustrates the component sheet metal fittings necessary to adapt the heat pump coil to the return air side of a counterflow furnace. Heat pump efficiency can be improved by insulating the return air duct in the attic space and by insulating the supply duct in the crawl space.

ADAPTING A COIL AND A COUNTERFLOW SYSTEM (COIL IN THE SUPPLY)

Figure 3-19 illustrates a counterflow furnace installation before the coil is installed. Figure 3-20 shows the coil and counterflow furnace after the heat pump coil is installed. Refer to Figs. 3-19 and 3-20 often while determining the fittings needed for your installation. The coil will be located between the furnace and supply

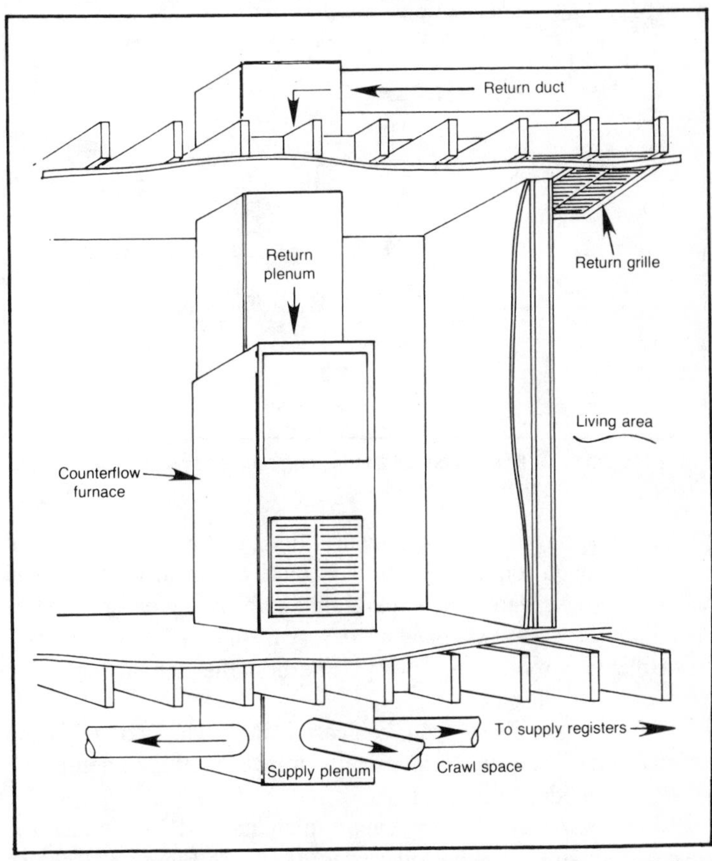

Fig. 3-19. Counterflow furnaces usually are used only on main floor installations. There are exceptions.

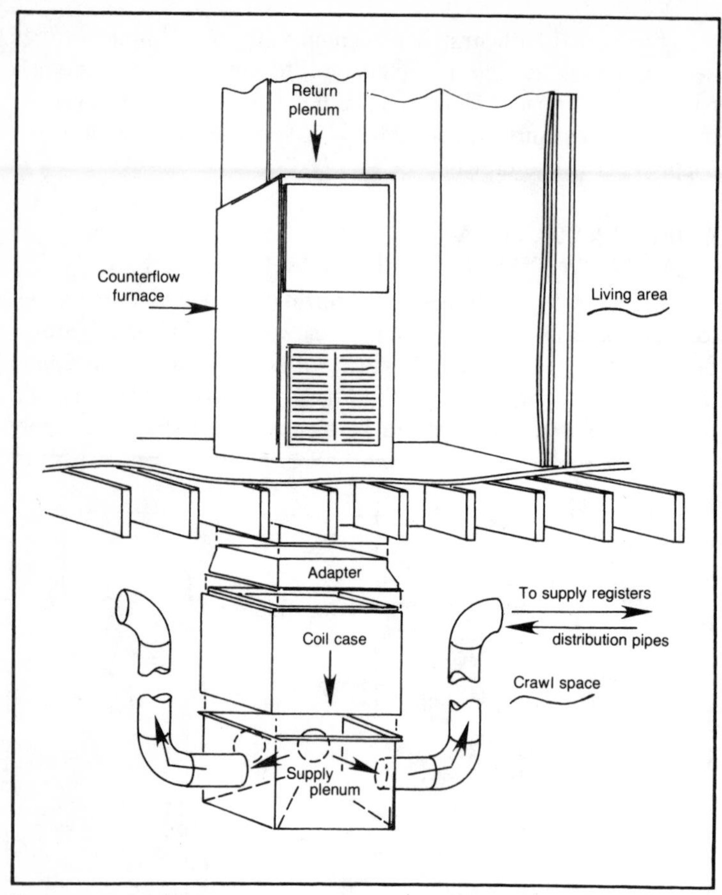

Fig. 3-20. Better air flow will be obtained if long radius turns are used to make changes of direction in the round pipe.

plenum. Looking at Fig. 3-20, the first thing that comes to mind is if there is enough room between floor joists and ground. If not, you will have to excavate. The excavation depth will be determined by the height of the coil case and necessary fittings. When excavating, leave plenty of room around the perimeter of the plenum to connect your distribution pipes.

Let's arbitrarily say your coil case is 26″ high. Taking this and other factors into consideration, let's draw a diagram to determine the space needed (Fig. 3-21).

Let's say your existing supply plenum is 19″ wide and 16″ deep. The inlet side of the coil case is 20″ × 20″. Figure 3-22 shows what you need for an adapter.

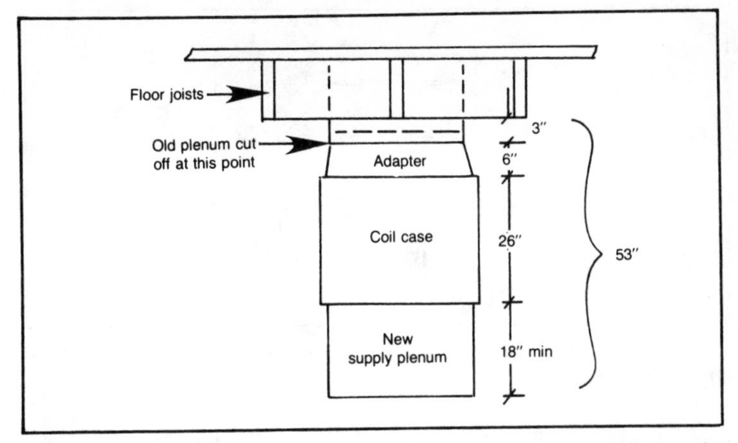

Fig. 3-21. Planning an installation is important when there is limited space for error.

To decide the size of the supply plenum, we must know the outlet size of the coil case. Let's say it is 19″ × 19″. Some coil cases have the same size inlet and outlet, and some do not. Let's assume this one doesn't. Figure 3-23 shows what is needed. Note the tapered bottom depicted by broken lines. This taper is to direct the air to the outside walls of the plenum, which will help the air flow toward the distribution pipes.

When you pick up your fittings at the shop, have the sheet metal mechanic explain how to make the seam connections between the adapter and old plenum cutoff point. He will probably recommend using S-clips and drives. If so, have him bend the drive pockets on fittings for you.

The adapter is installed first. If S-clips and drives are used, you also will have to bend the drive pockets on the appropriate sides of the old plenum to receive the similar pockets on the adapter. By

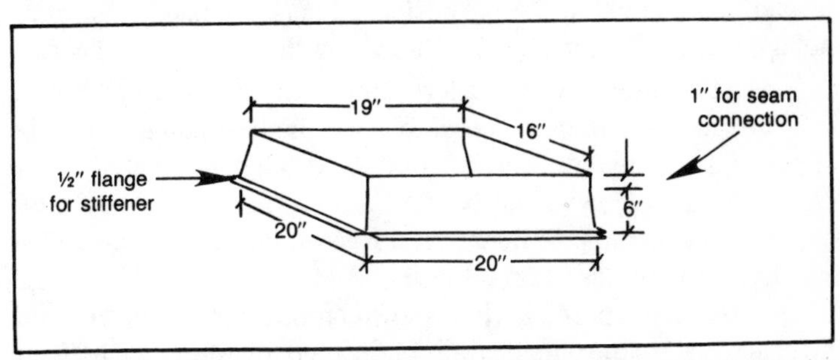

Fig. 3-22. If you design your own adapter, include enough room for connecting the adapter to other fittings.

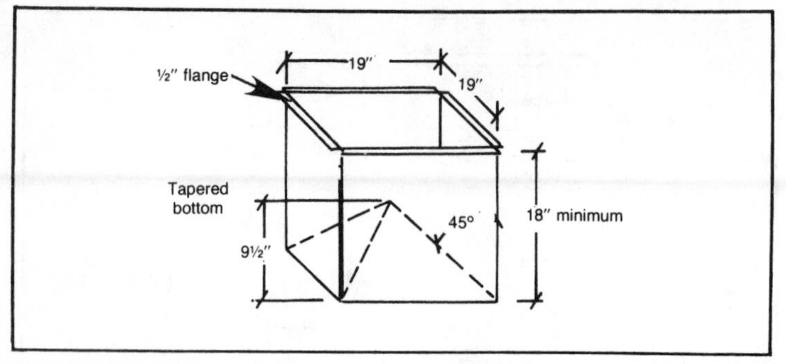

Fig. 3-23. If possible, adapters should be tapered evenly on all sides to promote uniform air flow.

following the directions received from the mechanic, you can attach the adapter to the old plenum cutoff point.

Run a bead of silicone caulking around the flange of the coil case and position the case up against the ½" flange of the adapter. You will need some help at this point to hold the coil until you can drill holes through the adapter and into the flange of the case to insert sheet metal screws. Be very careful that you don't drill into the coil or refrigeration lines.

When the coil case is secured, install the new plenum. Run a bead of silicone on top of the ½" flange and center the plenum over the outlet side of the coil case. Drill ⅛" holes through the flange into the case. Don't drill into the coil or refrigeration lines. Insert #7 × ½" screws.

Note that the distribution pipes are at a much higher point than the supply plenum. You will have to measure what is needed and purchase the appropriate fittings. Figure 3-20 illustrates the elbows in the distribution pipes.

All air duct in an unconditioned area should be wrapped with insulation. A crawl space is considered unconditioned. Insulation helps keep conditioned air from escaping through the air duct.

Most counterflow systems resemble that in Fig. 3-19. Figure 3-20 details the fittings required to adapt the heat pump coil to the supply side of the unit. Because of the extra length created by the extra fittings, elbows must be installed on the round distribution pipes to get them up to floor level. Excavating is required unless you have a very deep crawl space.

Notice the tapered bottom shown on the supply plenum. This directs the air to the outside walls of the plenum, which will help the air flow toward the distribution pipes.

ADAPTING A COIL AND A COUNTERFLOW SYSTEM (COIL IN THE RETURN IN THE ATTIC)

Figure 3-24 shows a typical counterflow system before a heat pump coil has been installed. Figure 3-25 illustrates counterflow system with the necessary fittings to adapt the coil to the unit. Refer to Figs. 3-24 and 3-25 often when you are determining the fittings needed for your installation.

The heat pump coil for this installation will be placed in the return air duct in the attic. Be certain you have an easy access to the attic area before starting this installation. The heat pump coil case is quite large and will take a considerable amount of space. It may require a modification of the existing access.

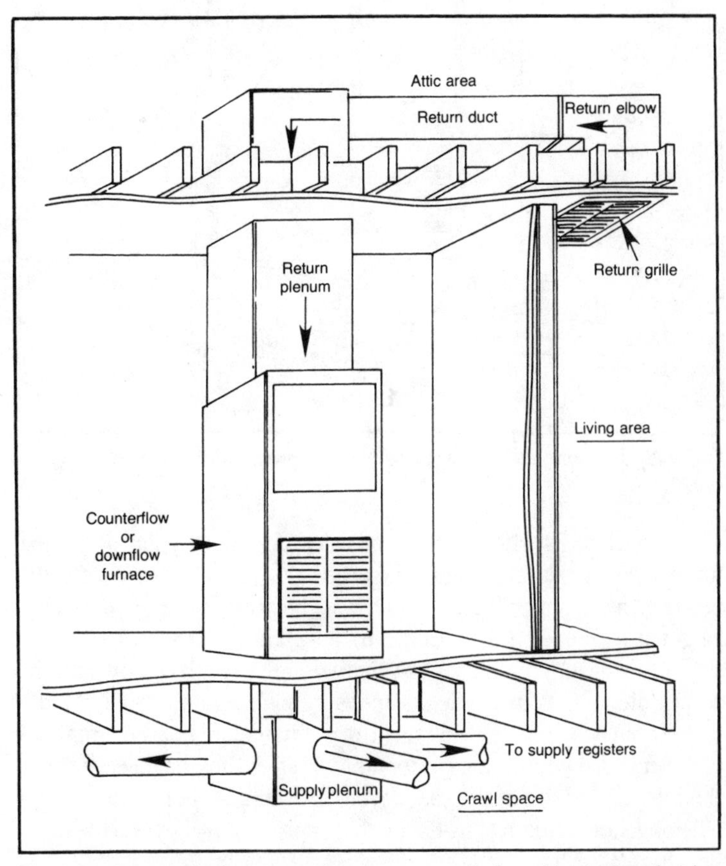

Fig. 3-24. Dust and other contaminants are usually found in ducting that has been in service for a few years. Precautions might be taken to ensure that the dust is controlled before working on the duct system.

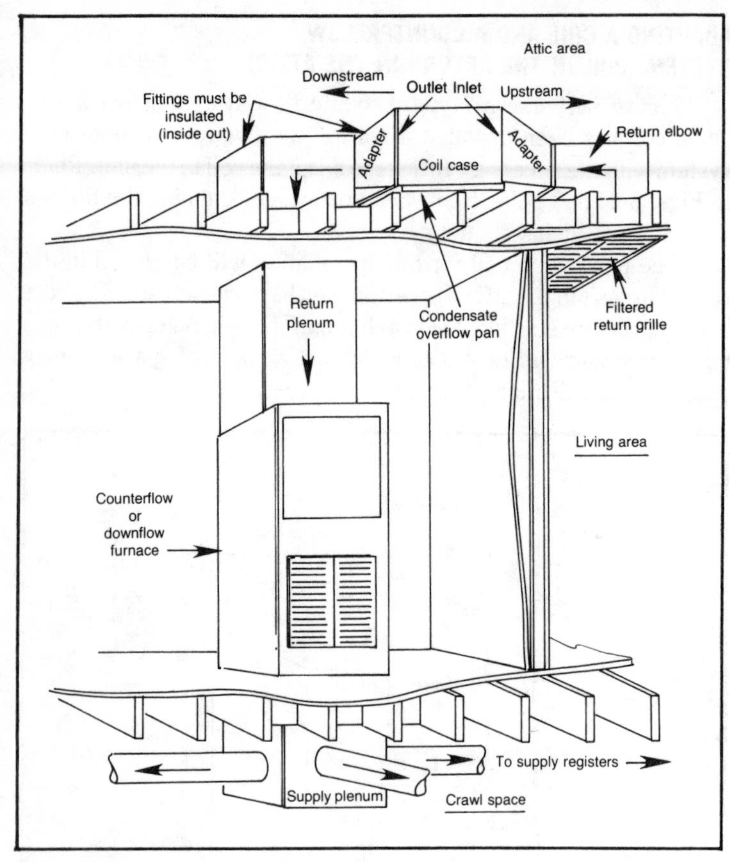

Fig. 3-25. Notice in this installation that the weight of the coil assembly is supported by the ceiling rafters.

Figure 3-24 should in some way resemble your particular furnace and air duct system. The area between the return air grille and return plenum will dictate the exact location of the coil case. Try to get it as close to the furnace as possible.

Let's assume you have 60″ between the return elbow and the return plenum. Your particular measurement will obviously be different, but we will use this arbitrary dimension as an example.

Measure your coil case length, which is the distance between the inlet and outlet of the case. We will assume that the coil case is 24″ long, not including any flange that is protruding beyond the case. See Fig. 3-26.

We have 60″ between the return elbow and return plenum. Subtract the 24″ length of the coil case from 60″ to get 36″. Let's

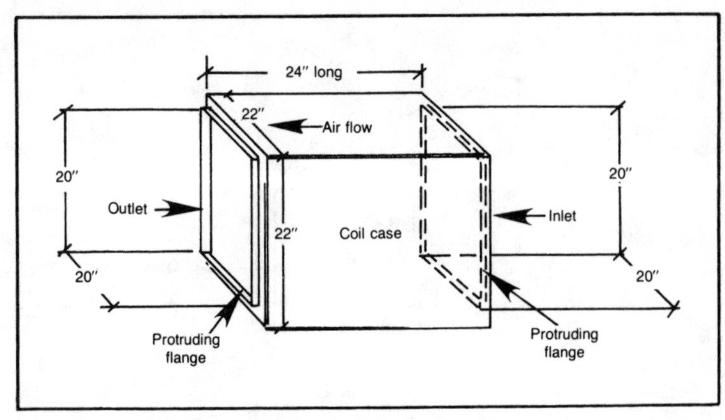

Fig. 3-26. Coil cases are sometimes attached to the duct material by S-clips, because they are manufactured with flanges.

plan on having the two adapters the same length to avoid confusion, so 36″ divided by two, equals 18″—not including our connection flanges. See Fig. 3-27.

The adapter on the inlet side of the coil case will be 19″ long, including the connection, and the adapter on the outlet side will be 18″ long. Let's assume the return air duct is 12″ deep and 20″ wide. Figure 3-28 illustrates how the two adapters will be fabricated.

Notice the condensate drain overflow pan under the coil case in Fig. 3-25. This pan is for emergency use in case the coil drain plugs up and the pan located within the case can't handle the excess condensation that melts off the coil. A drain must be installed on the bottom of the auxiliary drain pan. Figure 3-29 describes the design of the drain pan.

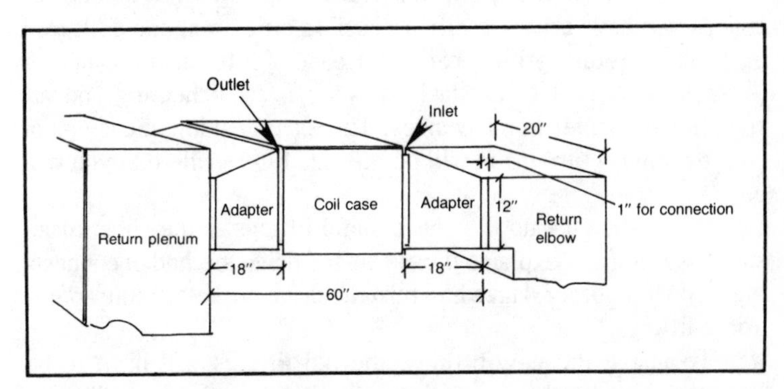

Fig. 3-27. Some type of support might be required if the duct system isn't strong enough to hold the weight of the coil and case.

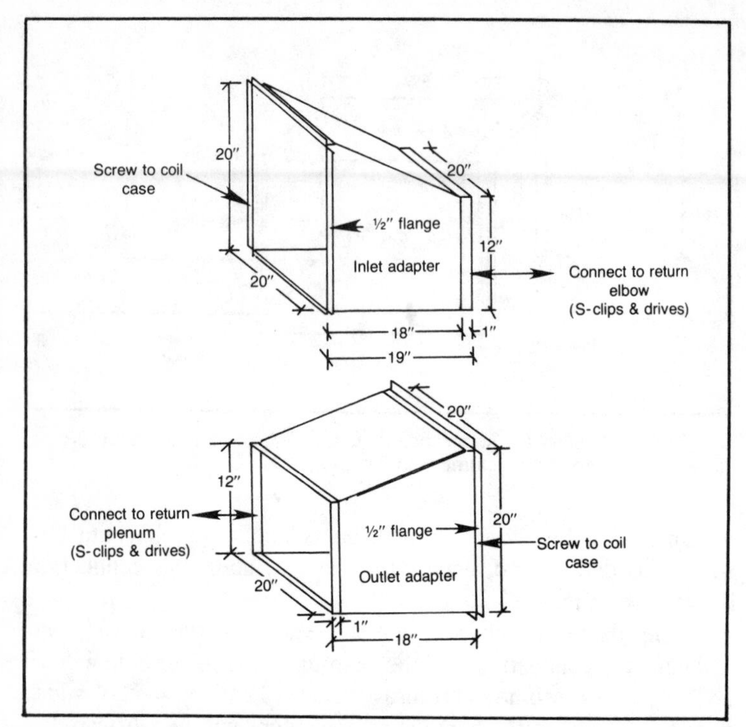

Fig. 3-28. A flange can be added to a custom outlet or inlet adapter. Your sheet metal shop will recommend the best method of attachment.

Note that the overflow pan is 4″ wider and longer than the coil case dimension. This is to allow for more capacity if needed.

The air filter of the system must be located upstream from the coil. The coil must be kept clean to enhance good air flow through it. Unless you have a very easily accessible attic area, use a filtered grille at the return grille location for ease of filter maintenance. A filter grille is readily available at your nearest heating and air conditioning dealer in many sizes. The exact inside dimensions of your existing return grille will be the size filter grille that you will need.

When you pick up the sheet metal fittings at the shop, make sure the mechanic explains the S-clip and drive method of connecting air duct together. Have him turn the drive pockets on the appropriate fittings.

To install the new fittings, the old fittings will have to be removed. Usually the connections are S-clips and drives. The information you received from the shop mechanic should be helpful in removing ductwork. In most cases a hammer, screwdriver, a pair of pliers, and a pair of aviation tin snips will do the job.

After the air duct has been removed, install the new fittings and the heat pump coil. The condensate overflow pan should be placed in the vicinity of where you plan to install the coil case. Set the pan on the ceiling joists and place the coil case in it. Refer to Fig. 3-25.

After the coil is set, install the two adapters to the coil case and existing air duct. The ½″ flange on the adapters will be screwed to the coil case using #7 × ½″ sheet metal screws. Be very careful when drilling into the coil case that you don't puncture the coil or the refrigeration tubing.

When the fittings and coil are installed, wrap all air duct that is downstream from the coil with insulation to avoid condensation forming on the outside of metal. Install the filter grille in place of the return grille.

Figure 3-25 shows the component sheet metal parts necessary to adapt the heat pump coil to the return air in the attic of a counterflow furnace. Compare Fig. 3-25 to Fig. 3-24. Notice the simplicity of the coil arrangement.

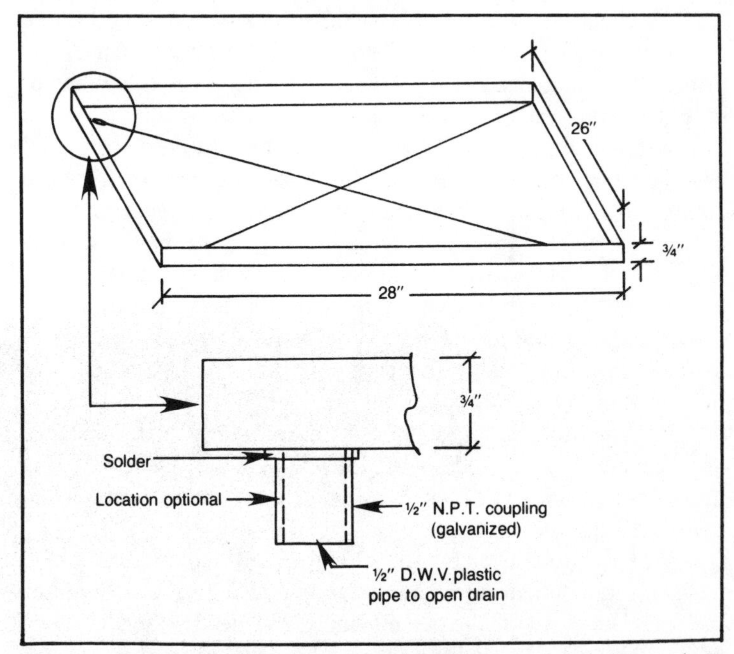

Fig. 3-29. If a condensate overflow condition develops, it might be wise to have an exterior emergency drain pan included in the installation. Exterior drain pans are found where building damage will occur if a water leak develops.

A condensate drain pan is located under the coil case in case the coil drain would plug up and overflow the drain pan located within the coil case. The pan must be piped to an open area for drainage if the auxiliary drain is needed for an overflow condition.

All air duct in the downstream side of the coil in an unconditioned area must be insulated to prevent condensation from forming on the air duct's metal surface. We recommend that all air duct in an unconditioned area be insulated. Your attic is an unconditioned area unless your insulation is installed on the roof structure. The crawl space is uninsulated unless the walls of the foundation are insulated with thick insulation.

ADAPTING A COIL TO A HORIZONTAL SYSTEM (COIL IN THE RETURN UNDER THE FLOOR)

Figure 3-30 shows a typical horizontal system before the heat pump coil is installed. Figure 3-31 illustrates a horizontal system with the necessary fittings to adapt the coil to the unit. Refer to Figs. 3-30 and 3-31 often when you are determining the fittings needed for your particular installation.

The heat pump coil in this installation is going to be located between the return connector elbow and the fan section of the furnace. The distance between these two points has to accommodate the fittings and the coil case.

Refer to Fig. 3-31. The required fittings, the furnace to coil case adapter, the filter and the filter rack, the adapter from the return elbow to the filter rack, and the coil case are shown. Three fittings are needed to install the coil. Figure 3-30 should in some way resemble your particular furnace and air duct system.

Measure your system from the outlet of the return connector elbow to the fan section inlet flange. This is the amount of space to install the coil case and the sheet metal fittings. Let's say you have 55″ between these two points.

Measure the length of the coil case, which is the distance between inlet and outlet. Suppose the length of the case is 24″. This measurement does not include any flange protruding beyond the case. See Fig. 3-32.

We have to use 3″ for the filter bracket. Now we have two unchangeable dimensions. The coil case is 24″ long, and the filter rack is 3″ long. Add these two dimensions to get 27″, and 27″ from 55″ equals 28″. Both adapters must be made up within the 28″ space. The connection is not included in any of these measurements. See Fig. 3-33.

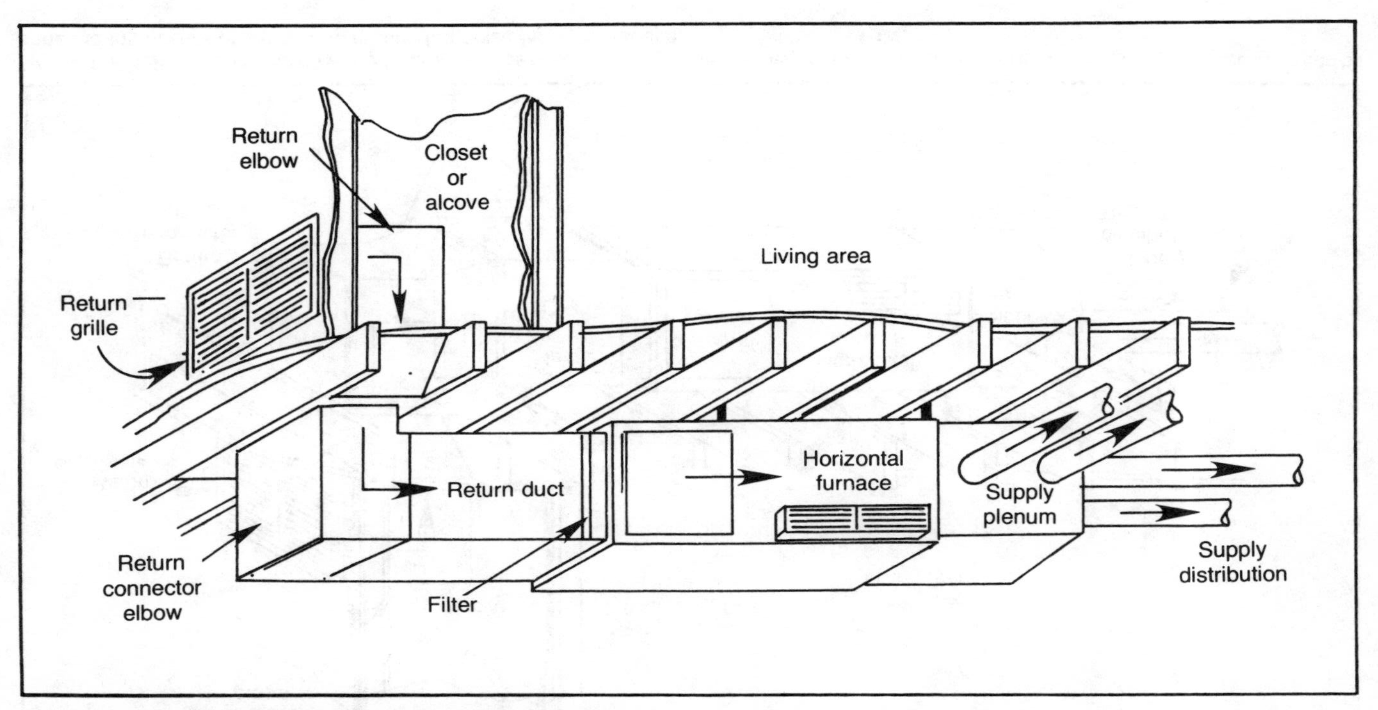

Fig. 3-30. Horizontal furnaces are used in areas of limited space.

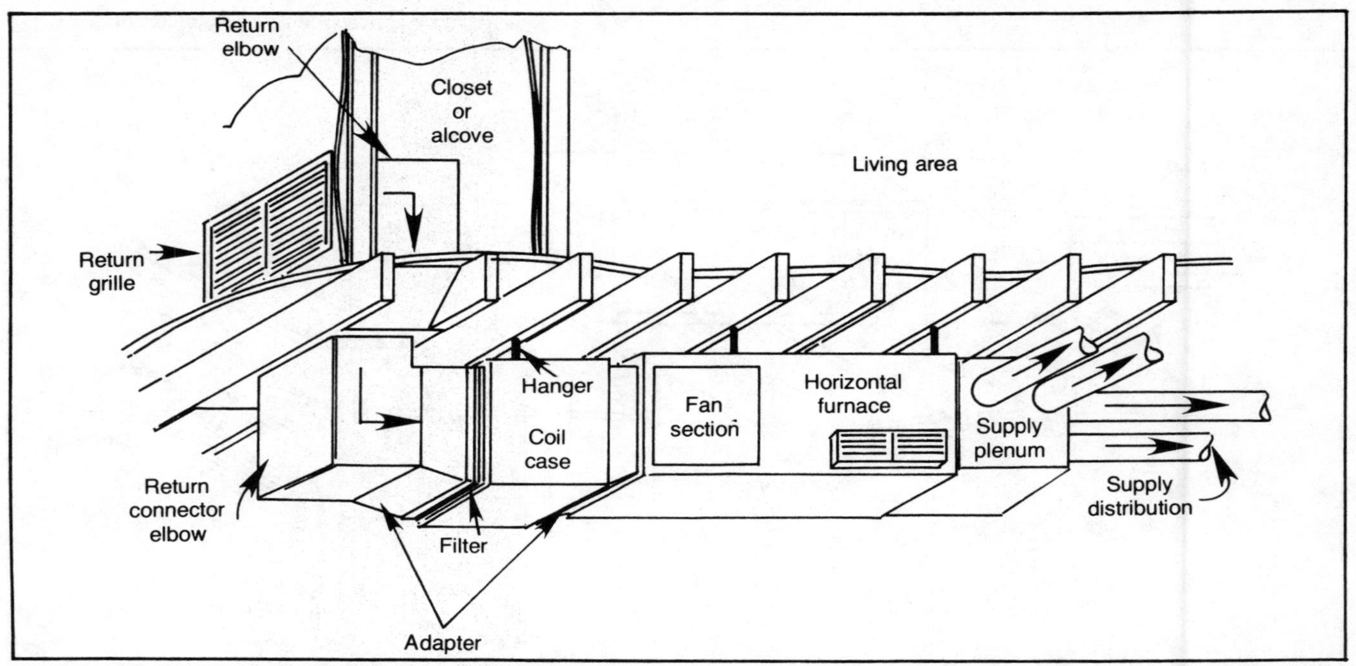

Fig. 3-31. Exterior duct insulation will reduce energy loss into an unconditioned crawl space. If the return duct is slightly oversized, insulation can be added to the interior of the duct to reduce fan noise from the furnace.

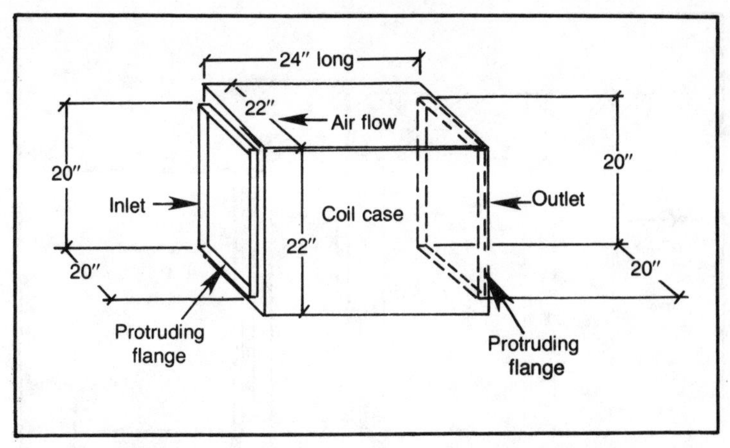

Fig. 3-32. Before planning an installation, be sure that you have a horizontal coil for a horizontal installation.

Note that the adapter on the inlet side of the coil is 14″ long plus 1″ on each end for the connection, which makes the total length of the adapter 16″. The adapter on the outlet side is 14″ long with a ½″ flange on each end that will be screwed to the coil case and the furnace. Figure 3-34 illustrates the furnace to coil adapter in detail. Let's assume the furnace fan inlet flange is 18″ high and 20″ wide.

The adapter attaches to the furnace fan section flange and to the coil flange. The filter rack attaches to the coil case on the inlet side with screws. The filter rack is detailed in Fig. 3-35.

When designing the size of the filter rack, be sure that the dimensions will accommodate a stock size filter. You can inquire at any hardware store for the particular sizes of air filters available.

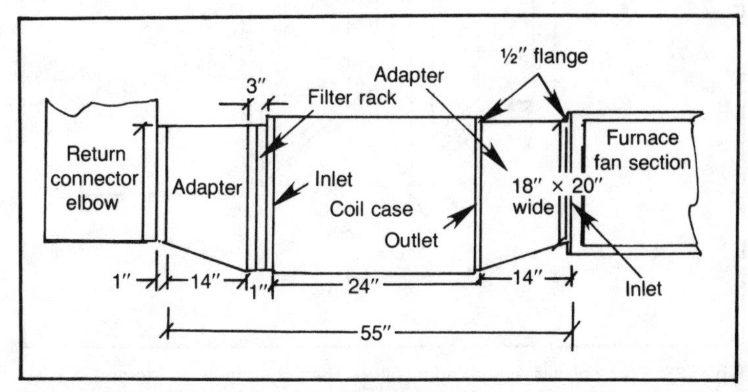

Fig. 3-33. Changing the size of an adapter should be done without any abrupt changes.

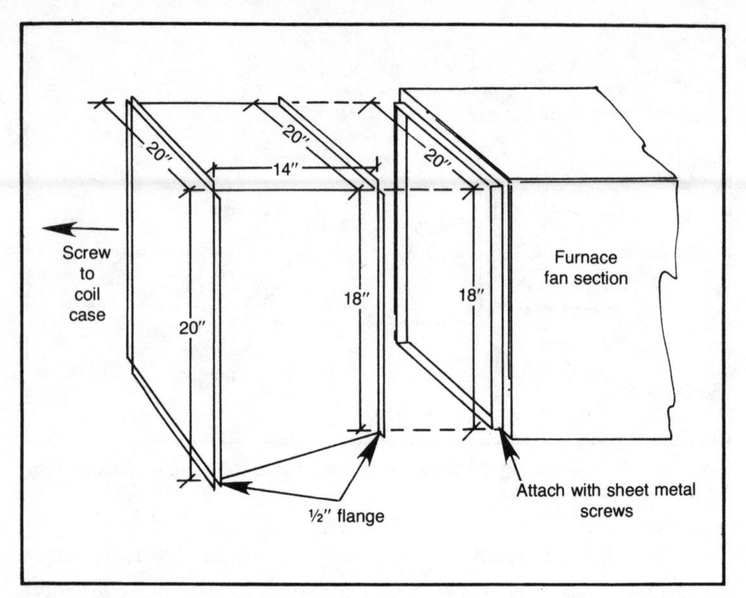

Fig. 3-34. Even though the sheet metal work may look complicated, it is easily explained by your sheet metal mechanic.

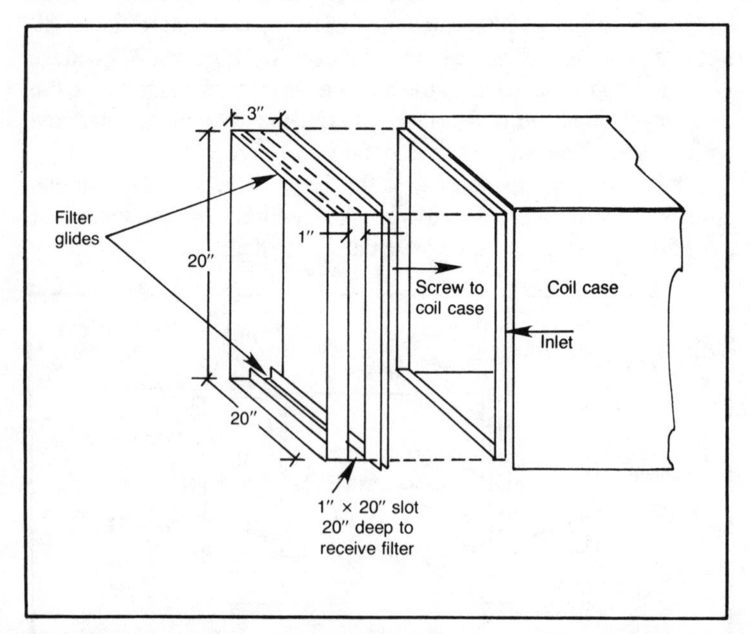

Fig. 3-35. Each coil will require an air filter—the larger the filter, the better. If we are moving 1,000 cubic feet of air per minute, a 2″ × 20″ × 25″ filter will be more effective than a 2″ × 16″ × 25″ filter. The size of the filter affects face velocity.

We will now design the adapter from the filter rack to the return connector elbow. Let's assume the outlet side of the elbow is the same size as the inlet side of the fan section, which is 18″ high and 20″ deep. This must be increased to fit the size of the filter rack, which is 20″ deep. This must be increased to fit the size of the filter rack, which is 20″ high and 20″ deep. See Fig. 3-36. This adapter will attach to both the filter rack and the return connector elbow with S-clips and drives.

When you pick up the fittings from the shop, ask the mechanic to explain the S-clip and drive method of connecting the air duct together. Have him put the drive pockets on the appropriate fittings.

To install the new fittings, the old ones will have to be removed. In most cases the connections will be S-clip and drive ones. The information you received from the sheet metal mechanic will be very helpful. You will understand the mechanics of the connection and will be able to disconnect it easily. In most cases a hammer, screwdriver, a pair of pliers, and a pair of aviation tin snips will do the job.

After the air duct has been removed, install the new fittings and heat pump coil. The first fitting to install is the adapter from the fan section to the coil case. Put a bead of silicone caulking around the outside of the inlet flange of the furnace's fan section. Place the adapter against the caulking. Insert #7 × ½″ sheet metal screws through the flange into the fan case to secure the adapter.

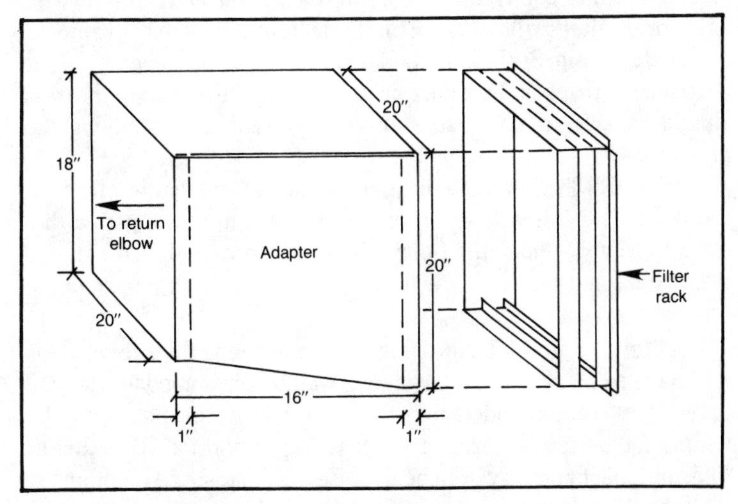

Fig. 3-36. A larger filter can be accommodated by the use of slightly larger adapters.

Install the coil case. You will need some means of support. If possible, get some extra help. Devise some type of hanger for the coil case and attach to the floor joist. Screw the adapter to the coil case. Be careful not to drill into any refrigeration lines, inside the coil case.

The air filter is attached to the coil case by screwing through the flange into the case. Be careful while drilling the holes.

After the filter rack is installed, connect the adapter from the filter rack to the return connector elbow. This is done by the S-clip and drive method. There is usually enough slack in the hangers of the furnace ducting, so you are able to force it enough to get the distance required, for the adapter to fit between the filter rack and the return connector elbow.

ADAPTING A COIL TO A HORIZONTAL SYSTEM (COIL IN THE RETURN ABOVE THE FLOOR)

Figure 3-37 shows a typical horizontal furnace before a heat pump coil is installed. Figure 3-38 illustrates a horizontal furnace and the necessary fittings to adapt the coil to the unit. Refer to Figs. 3-37 and 3-38 often when determining the fittings needed for your particular installation.

The heat pump coil for this installation will be located on the main floor in a concealed area such as a closet or alcove. The return air drop will have to be extended to allow space for the coil. The return grille has been moved higher up on the wall and replaced with a return air filter grille to make filter maintenance more convenient.

Refer to Fig. 3-37. This should in some respects resemble your particular furnace and air duct system. With a few exceptions, you should be able to visualize your own system after studying the diagram.

Figure 3-38 illustrates the method used to install the coil above the floor. The coil will sit on the floor in a waterproof pan with a drain to an area under the floor. The pan is in case the coil should ever freeze and cause an overflow of water that the drain pan in the case couldn't handle.

Refer to Fig. 3-37. Notice the return grille at floor level. This will have to be raised higher on the wall to accommodate the coil case. It is recommended that the return grille be placed at the ceiling for better air circulation. It is very important that the air filter on a heat pump system be kept free of dust, so we recommend a filter grille to replace the return grille. A filter grille contains an air filter with a hinged front that is easily opened for filter removal.

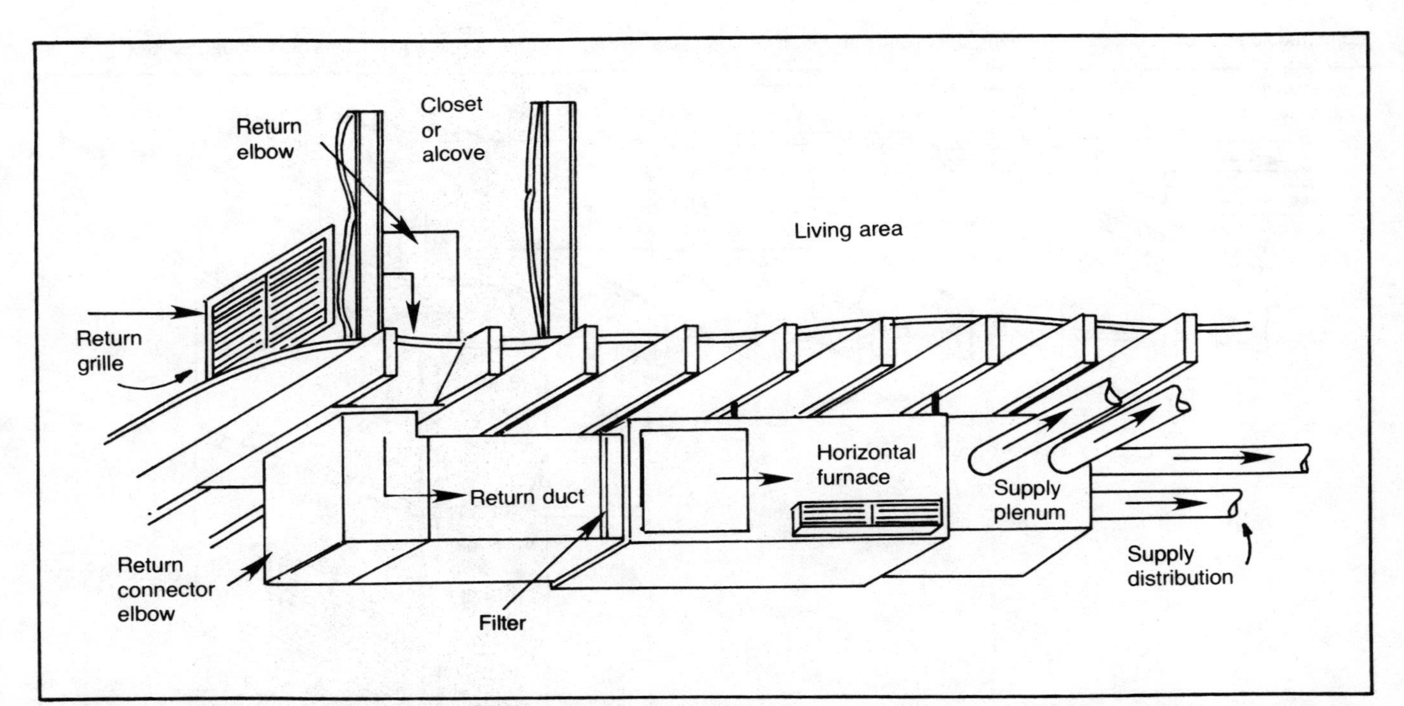

Fig. 3-37. If a horizontal furnace is supported from the ground instead of the floor joists, quieter operation will be attained.

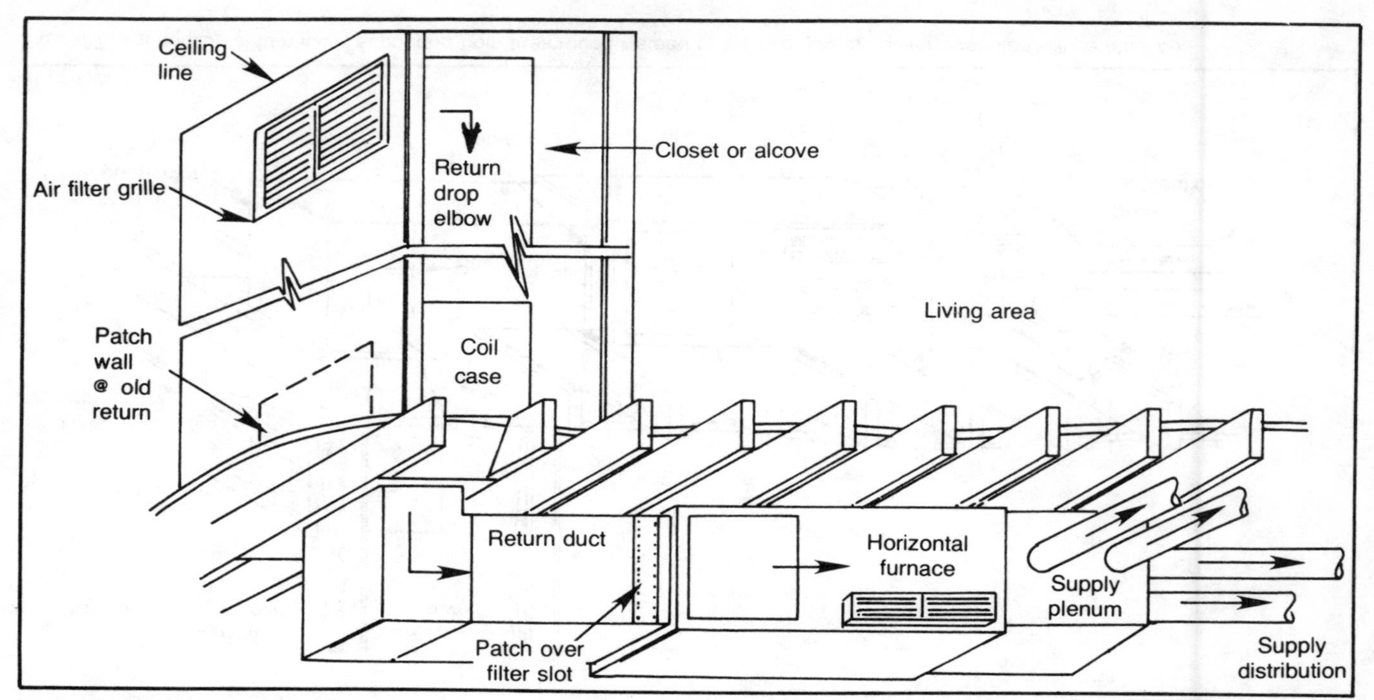

Fig. 3-38. When changing the location of a return air grille, the sheet wall can be easily patched using Sheetrock, tape, and spackling compound.

Filter grilles come in many sizes, so you should have little trouble finding one that will fit your need. An oversized grille is much more desirable than a smaller one in that it cuts down on air velocity noise and also delivers a better volume of air across the coil, which will enhance the performance of the heat pump.

Figure 3-39 gives you an idea of a typical coil installation, and with a little imagination it can be converted into your particular need. Figure 3-39 has more detail on the design that will be used.

Let's describe each fitting needed for this particular installation. We will start with the waterproof pan as it attaches to the outlet of the coil case. We will hypothetically say the coil case is 24″ high, and the outside of the case is 22″ square. The inlet and outlet of the coil case are both 20″ square. Figure 3-40 will indicate what is needed for the drain pan.

Note that the center of the drain pan is opened to allow air flow from the outlet side of the coil case to the return air on the furnace. The outside of the coil case is 22″ square, and the outside of the drain pan is 26″ square. This allows space around the edge for water drainage. A drain line must also be installed to the drain pan. This secondary pan is for emergency purposes only. The chances of a coil fan ever overflowing into it are very slim.

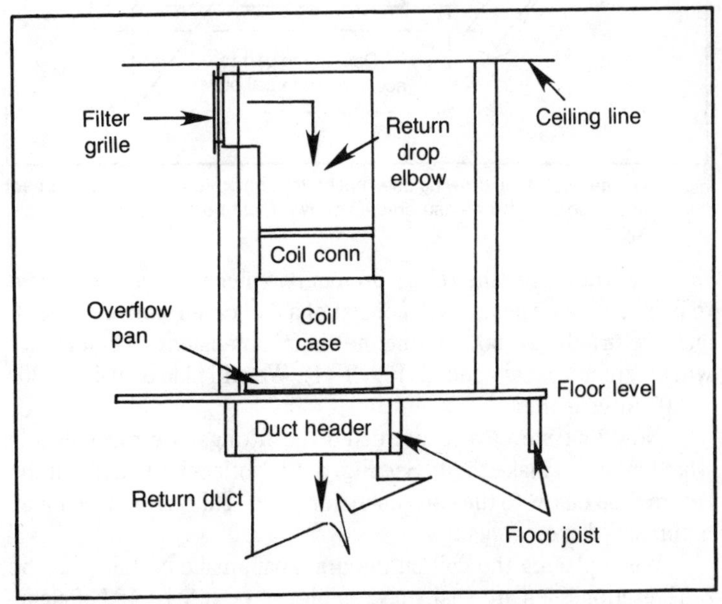

Fig. 3-39. A high wall return air grille may be an advantage in your installation. Preplanning will make the project much easier.

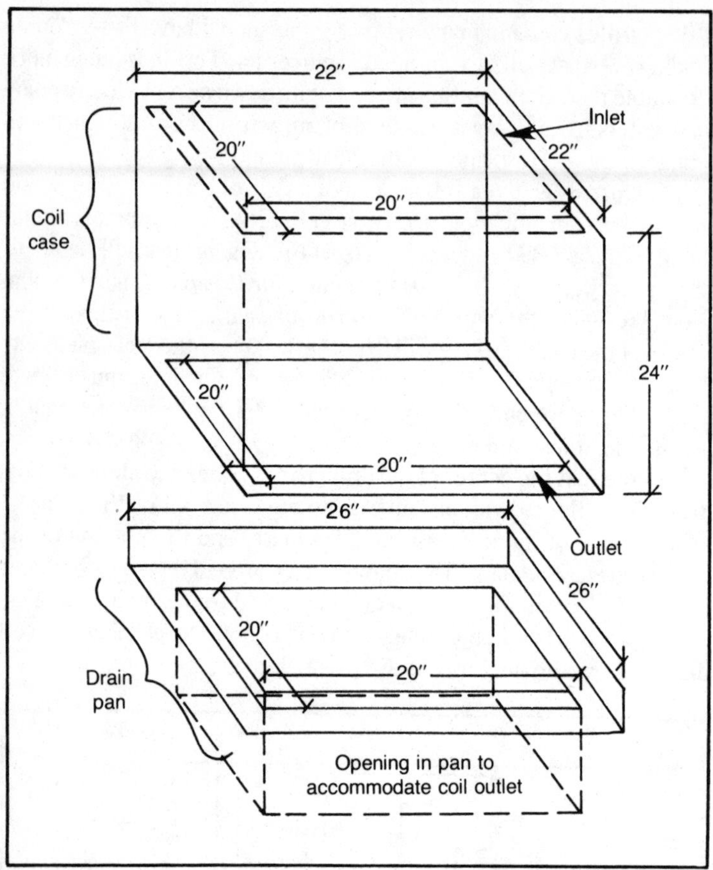

Fig. 3-40. The duct fitting drawing does not have to be professional for your sheet metal mechanic, but the measurements shown must be accurate.

We will design the return drop in two pieces, which enables it to be installed easier. It will consist of a coil case return connector and a return drop elbow. Using the same dimensions as before, we will diagram what is needed (Fig. 3-41). We will plan on using a 20″ × 20″ filter grille.

Now that we have constructed the fittings, we must decide what lengths to make them. See Fig. 3-42. Notice the 1″ extra at the return drop elbow to the coil connector seam. This is required for an S-clip and drive connection.

When placing the coil on the drain pan, make certain that the refrigeration lines are easily accessible. Have at least 18″ of clearance for installation of the refrigeration live set. Also, have the access door facing toward the opening of the alcove as the coil needs periodic cleaning.

You can compose a complete illustrated fitting list that can be taken to a sheet metal shop for fabrication. When you pick up the fittings at the shop, have the mechanic explain the S-clip and drive method of connecting the air duct. Also, have him turn the drive pockets on the appropriate pieces, which will be the return drop elbow and the coil connector.

Before you start installing the coil and fittings, it will be necessary to find a place for the hole that you must cut in the floor above the return air duct. It will be cut to the size of the coil case outlet size and the drain pan opening. After the hole is cut for the outlet of the coil and drain pan, locate and drill a hole large enough to accept the drain from the pan. This is for emergency purposes only and will not require a receptacle for the water; let it drain to the soil under the floor.

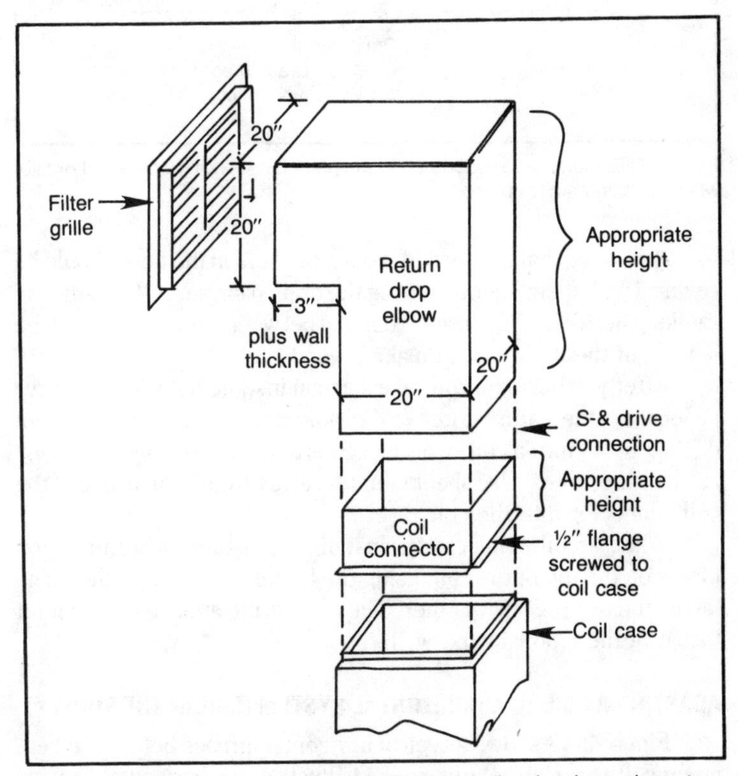

Fig. 3-41. An exploded view of a project often helps in the planning and construction stages.

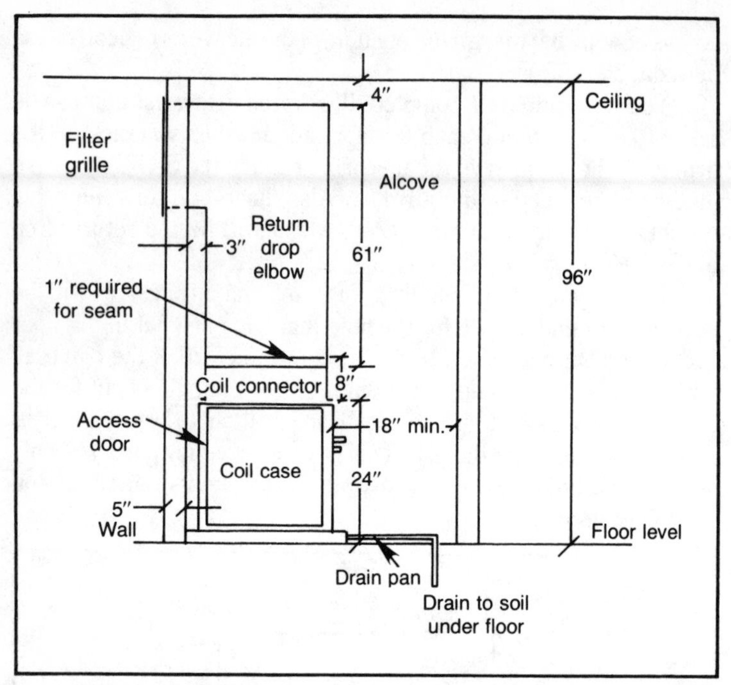

Fig. 3-42. Remember to include enough space for servicing of the coil or filter when installing a heat pump.

Any wood that is exposed around the hole in the floor should be covered with metal before setting the drain pan over it. Use silicone caulking between the pan and floor and between the coil case and the bottom of the drain pan to make a good seal.

After you have the coil case and pan installed, it's just a matter of locating the return filter grille hole and cutting it out. When drilling holes into a coil case, make certain you don't puncture any refrigeration lines. Use sheet metal screws through the flange of the coil connector into the coil case.

When a coil connector is installed, position the return drop elbow on the top of the coil, using the S-clip and drive method, and fasten the two pieces together. The filter grille attaches to the inlet throat of the elbow at the wall.

ADAPTING A COIL TO A HORIZONTAL SYSTEM (COIL IN THE SUPPLY)

Figure 3-43 shows a typical horizontal furnace before the heat pump coil is installed. Figure 3-44 illustrates a horizontal furnace and the necessary fittings to adapt the coil to the unit. Refer to Figs.

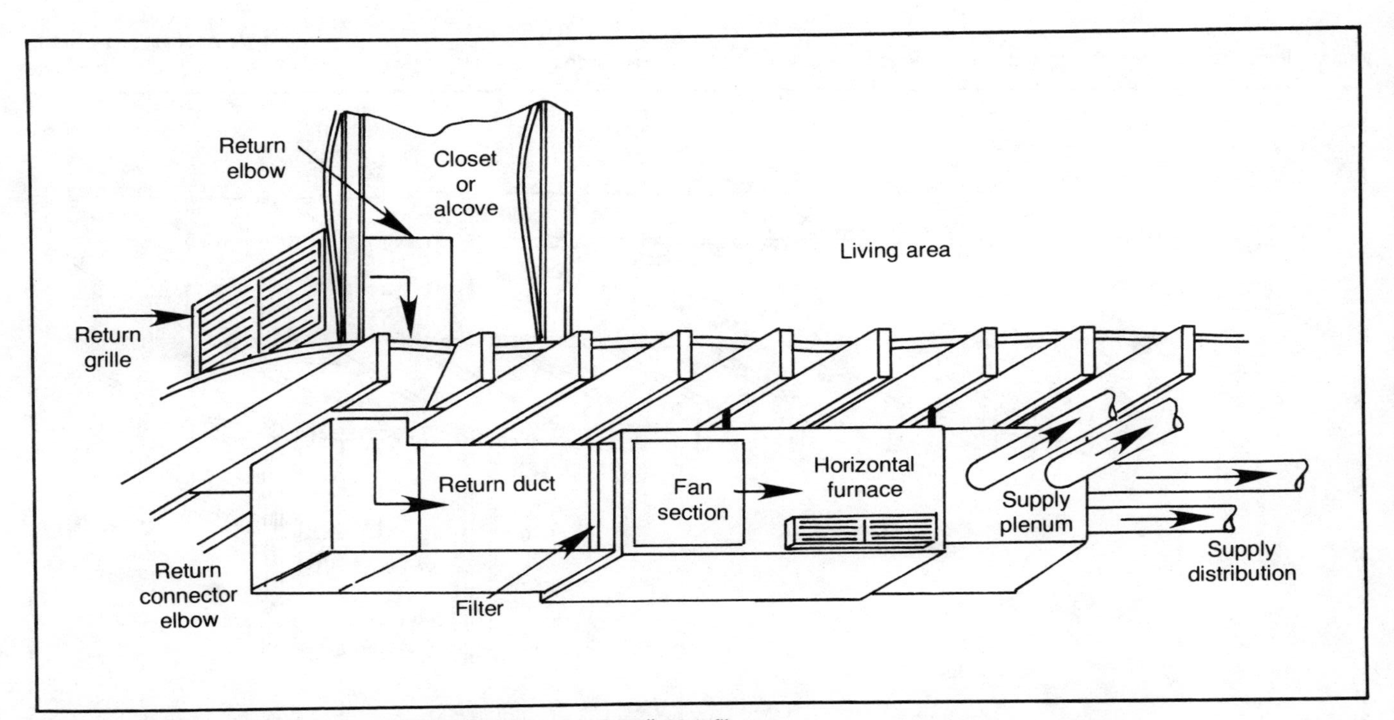

Fig. 3-43. Whenever possible, use a 2″ thick filter instead of a 1″ thick filter.

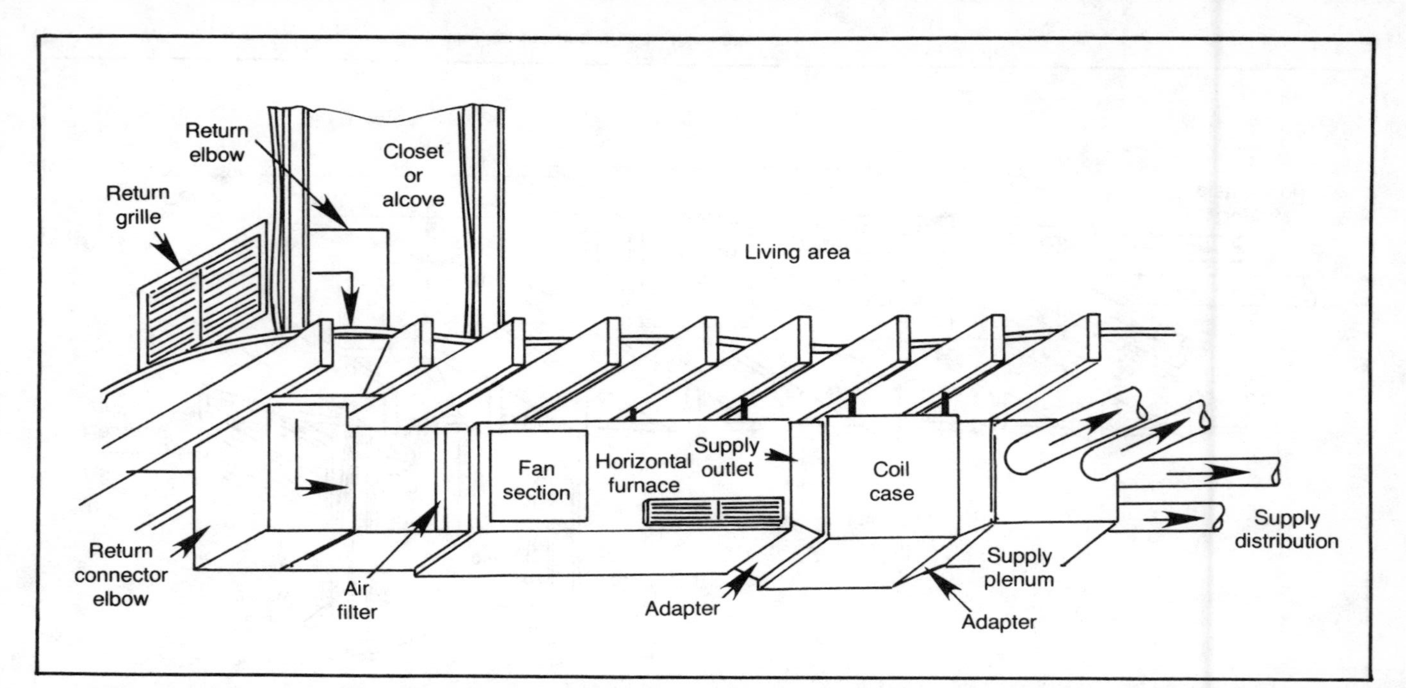

Fig. 3-44. When the coil of a heat pump is installed in the supply air duct, special controls must be used so that the furnace and heat pump can not operate simultaneously.

3-43 and 3-44 often while you are determining the fittings needed for your particular installation.

The heat pump coil for this installation is going to be located between the furnace supply outlet and the supply plenum. Notice that this furnace is hung from floor joists. This is a typical procedure with a horizontal furnace, but it is not always used. Some furnaces are set on cement blocks. You will have to devise a means of support for the coil case when you are ready for installation.

The required fittings are shown in Fig. 3-44. Two fittings are needed to install the coil. The coil case is generally larger than the outlet of the furnace and supply plenum, which will require an adapter from one size to the other to accommodate the differences. Two adapters or transitions are needed. Figure 3-43 should in some way resemble your particular furnace and air duct system.

Before starting your installation, make certain you have a large enough access under your floor to accommodate the heat pump coil. The first measurements we will need are the size of the supply outlet of the furnace and the size of the inlet of the coil case opening. Let's assume the supply outlet is 20″ high and 18″ wide, and the coil case inlet is 24″ high and 20″ wide. Figure 3-45 illustrates the necessary measurements.

When determining the adapter size, decide the length of the fitting also. To accommodate the length of the coil case, the supply plenum will have to be moved forward. We must decide at this time what distance can be tolerated.

Let's assume the coil case is 24″ long, and we decide to have the adapters 6″ long. These three figures add up to 36″. Make sure

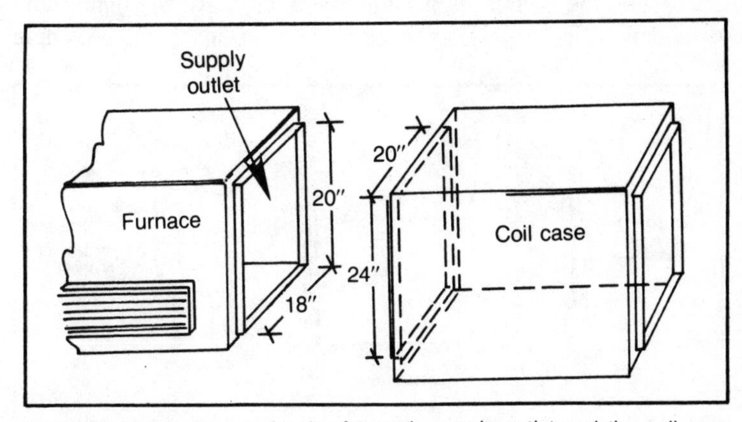

Fig. 3-45. Measurements for the furnace's supply outlet and the coil case opening's inlet.

the supply plenum can be moved this distance without creating a big problem. Assuming this can be accomplished, we will now illustrate the design of the two adapters (Fig. 3-46).

The plenum was screwed to the furnace, so we will also assume the plenum is the same size as the outlet side of the furnace. The outlet of the coil is also generally the same size as the inlet. Figure 3-46 depicts both adapters.

As mentioned above, the plenum will have to be moved ahead 36″ to accommodate the fittings and the coil case. You will have to change the lengths of the distribution pipes, which may require some extra pipe and fittings.

To install the fittings, you will need an electric drill with an ⅛″ bit, some #7 × ½″ sheet metal screws, a screwdriver or screw chuck for the drill, a hammer, and pliers. Remove the existing supply plenum. See how it is attached and remove the distribution pipes and plenum.

The first fitting to install will be the adapter between the furnace and the coil. Place the appropriate end of the adapter over the flange of the furnace and screw it to the flange. Position the coil against the outlet end of the adapter. Drill holes through the adapter into the flange of the coil case. Be careful not to drill into the coil or refrigeration lines. You will need some means of support for the coil case. You can use some blocks to position it and add hangers after it is secured. When the coil case has been secured, install the other adapter to the outlet end of the coil using the same method as above.

The plenum can be screwed to the adapter through the flanges. See Fig. 3-47.

When the plenum is installed, you will have to connect the distribution pipes to it. With the plenum in position, it will be simple

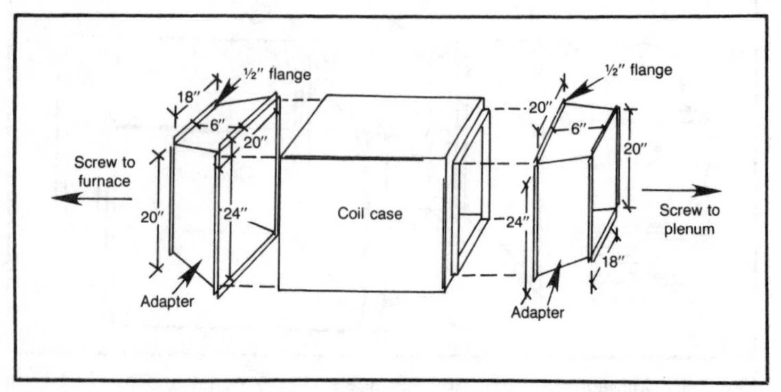

Fig. 3-46. Rarely will a coil case ever fit a furnace or existing ductwork.

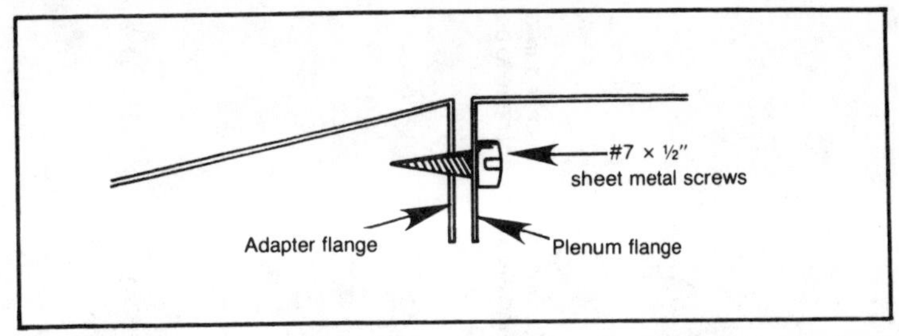

Fig. 3-47. Sheet metal flanges can be attached on the inside of the ducting if space permits.

to measure for the extra pipe needed. You can purchase most any size pipe you need from a sheet metal shop. Install the distribution pipes and the coil as needed.

ADAPTING A COIL TO A LOWBOY SYSTEM (COIL IN THE RETURN)

Figure 3-48 shows a typical lowboy furnace before the heat pump coil is installed. Figure 3-49 illustrates a lowboy furnace and the necessary fittings to adapt the coil to the unit. Refer to the Figs. 3-48 and 3-49 often while you are determining the fittings needed for your particular installation.

The heat pump coil is going to be installed between the return duct and the fan section of the furnace. Because of the design of the lowboy furnace, it is relatively easy to adapt a coil to it. The inlet of the fan case has a ready-made base for the coil to set on. All that's required is usually transition from the size of the coil case to the inlet of the return side of the furnace.

Figure 3-49 shows the required fittings, a flex connector, a filter rack, the coil case, and the adapter or transition. The coil is in the return air side of the furnace, which is acceptable in an area of low humidity. Figure 3-48 should in some way resemble your own particular furnace and duct system.

The first measurement needed is the height of the return drop. We will hypothetically say the measurement is 42″. This is the amount of space to install the heat pump coil and the fittings. Out of the four items needed in the coil connection, the three with unchangeable measurements are the coil, the filter frame, and the flex connector. The flex connector is manufactured in 6″ width, and at least 3″ for the filter frame and the height of the coil case are needed. Let's say that the coil case is 24″ high. Using all the information above, we can draw a diagram of what is needed as far as height is concerned (Fig. 3-50).

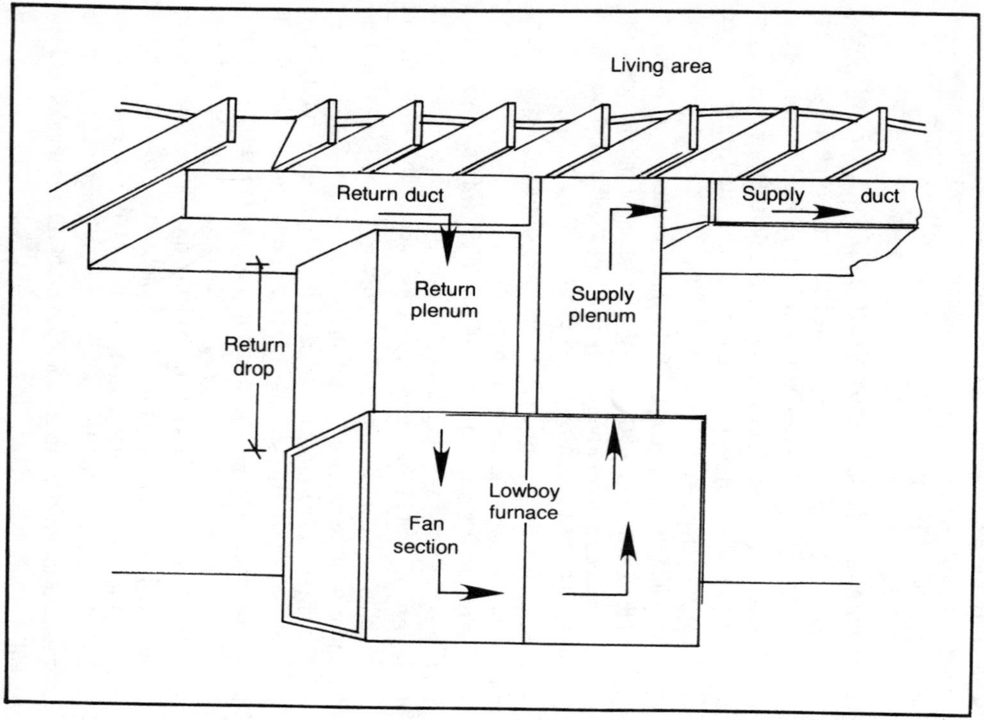

Fig. 3-48. Duct that is installed in a conditioned area does not have to be insulated.

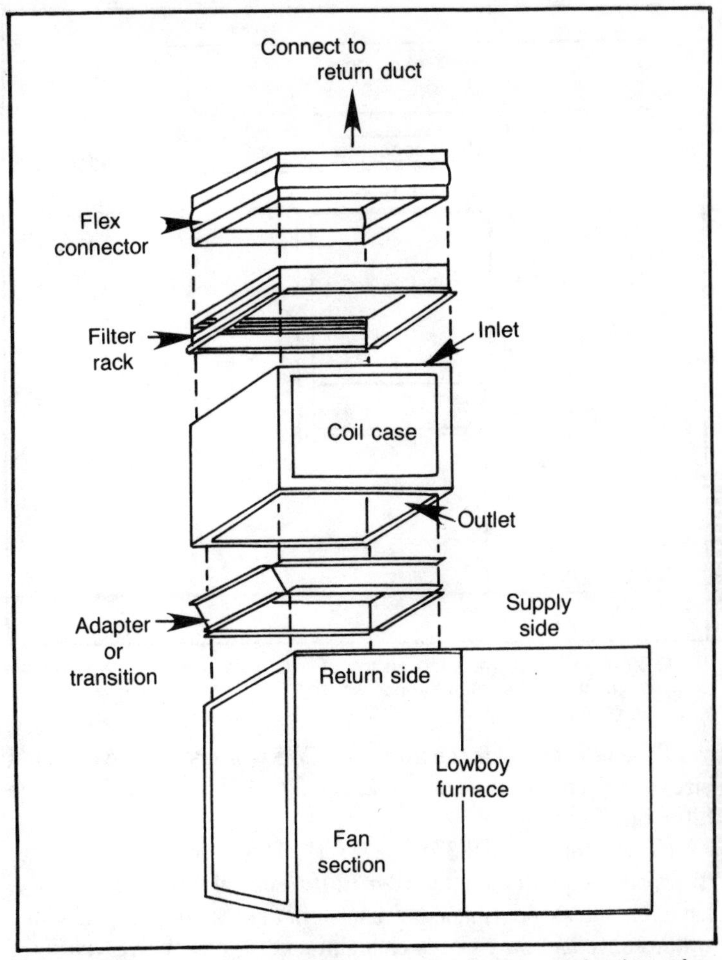

Fig. 3-49. Increasing air filter efficiency is as simple as increasing the surface area exposed to air flow within the duct. By increasing the surface area, the speed that the air moves through the filter material is reduced, thus giving the filter a better opportunity to trap duct particles.

The flex connector and the filter rack measurements equal 9″. The S-clip and drive connection will take up 1″, so this leaves us with a net of 8″ from the return duct to the inlet side of the coil case.

Now that we have the height of each fitting, let's decide what we will need for the opening sizes for each. We will start with the adapter between the outlet of the coil and the furnace. Let's say the return flange on the furnace is 20″ wide and 20″ deep and the outlet side of the coil case is 22″ wide and 22″ deep. See Fig. 3-51.

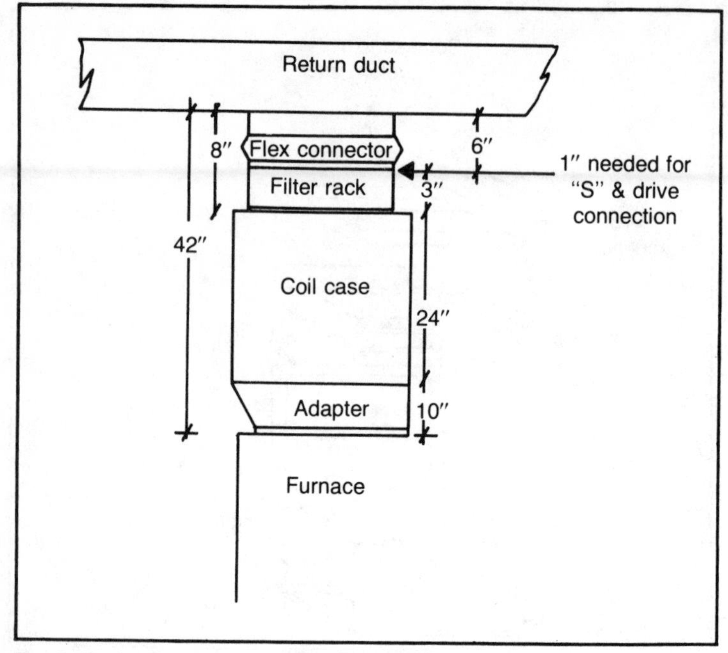

Fig. 3-50. A coil case mounted on top of a furnace usually doesn't require additional support other than the ductwork.

The next fitting is the filter rack. We will assume the coil inlet size is the same as the outlet—22″ × 22″. We will now design the filter rack (Fig. 3-52).

The filter will be 22″ × 22″ × 1″. This is not a standard size filter, and you will have to order it. If a special filter is necessary, have it made as a permanent reusable one. It can be cleaned with a water spray and only needs changing every four or five years.

The flex connector will fit the inlet side of the filter rack. It will be the same size in diameter. See Fig. 3-53. The flex connector fastens to the return duct and the filter rack with S-clips and drives.

Compose a complete illustrated fitting list that you can take to a local sheet metal shop and have the fittings fabricated. Before the coil and fittings can be installed, remove the old return drop. You can generally remove the screws from the flange at the furnace and the S-clips and drives at the return duct.

The first fitting to install will be the adapter between the coil and the furnace. Using #7 × ½″ sheet metal screws, attach the appropriate opening to the outlet side of the coil case. When drilling into the coil case, don't drill into the refrigeration lines or coil.

After the adapter is secured to the coil case, place the coil and adapter onto the return flange on the furnace and secure with screws. See Fig. 3-49.

The filter rack is attached to the top of the coil case with sheet metal screws through the flange. Again, be careful not to drill into refrigeration lines or the coil.

The last fitting is the flex connector, which makes it a lot easier to connect to the two fittings. It would be a real chore with a solid fitting. The flex connector is installed using the S-clip and drive method.

ADAPTING A COIL TO A LOWBOY SYSTEM (COIL IN THE SUPPLY)

Figure 3-54 shows a typical lowboy furnace before a heat pump coil is installed. Figure 3-55 illustrates a lowboy furnace and the

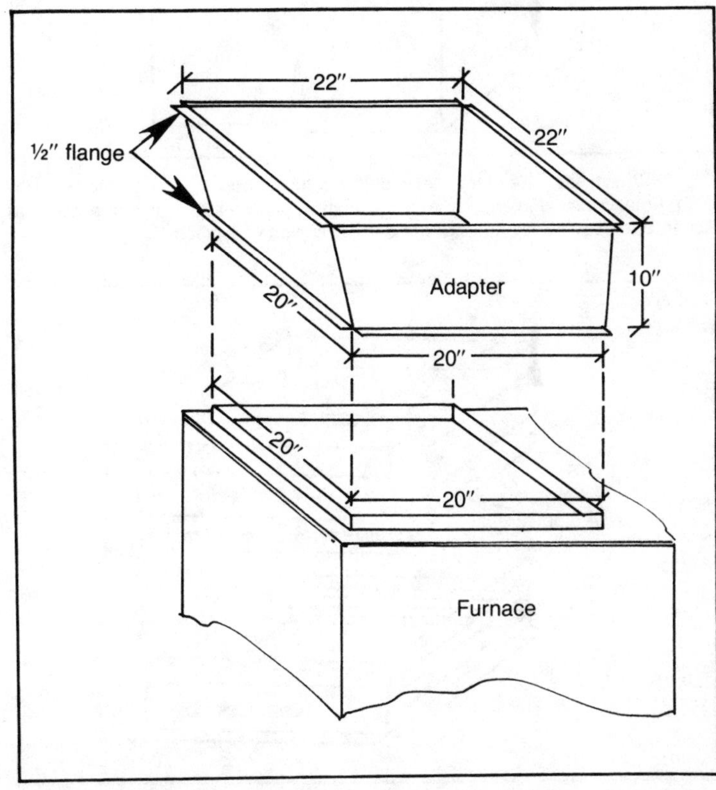

Fig. 3-51. Duct adapters can be designed in almost any configuration to meet a specific air flow pattern. The best approach to a design problem is usually the simplest method available.

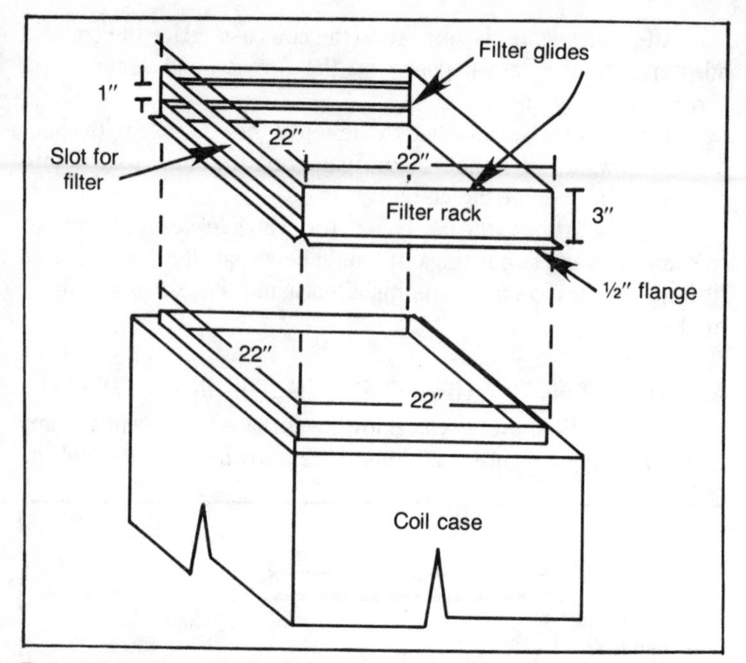

Fig. 3-52. Air filter maintenance should be performed on a monthly basis to obtain maximum efficiency from your heating plant. Note that in the diagram above each section of ductwork is designed piece by piece.

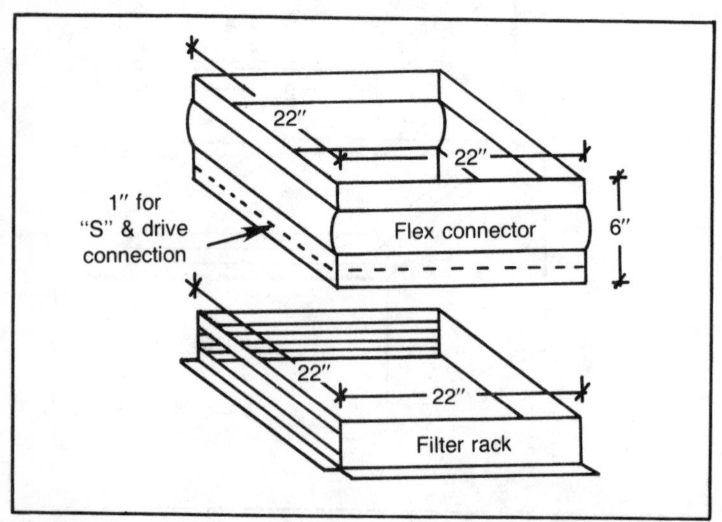

Fig. 3-53. A flex connector can be installed in any section of duct near the furnace or fan assembly. Attach each section of ducting by using the method recommended by your sheet metal shop.

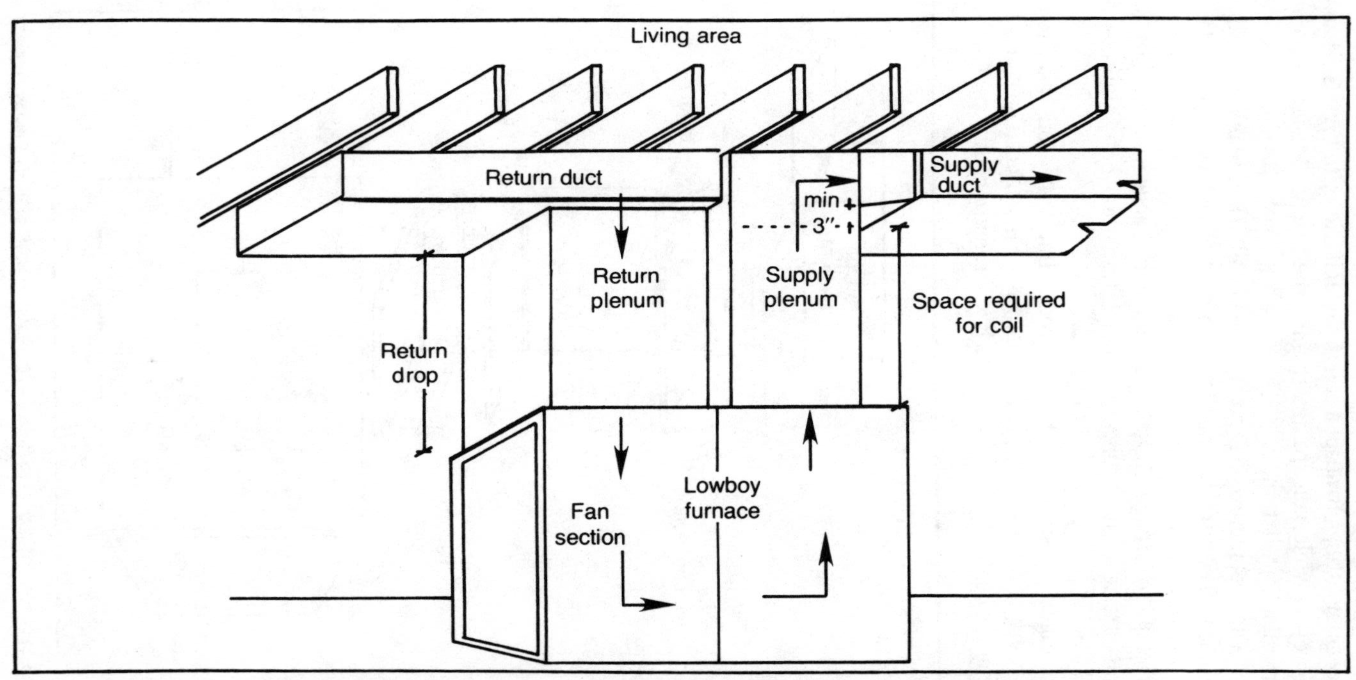

Fig. 3-54. All duct material is attached to the ceiling of the basement in this particular installation. A lowboy furnace is found in many older installations.

necessary fittings to adapt a coil to the unit. Refer to Figs. 3-54 and 3-55 often while you are determining the fittings needed for your particular installation.

The heat pump coil is going to be installed in the supply plenum area. It will require cutting the plenum off enough to accommodate the coil and fittings. The coil is generally larger than the supply outlet of the furnace. Adapters are used to change from one size to another size. This fitting is called a transition or sometimes a reducer.

Figure 3-55 shows the required fittings, the coil case, two adapters, and the supply plenum that will be modified. The coil is in the supply side of the system, which will require cutting part of the

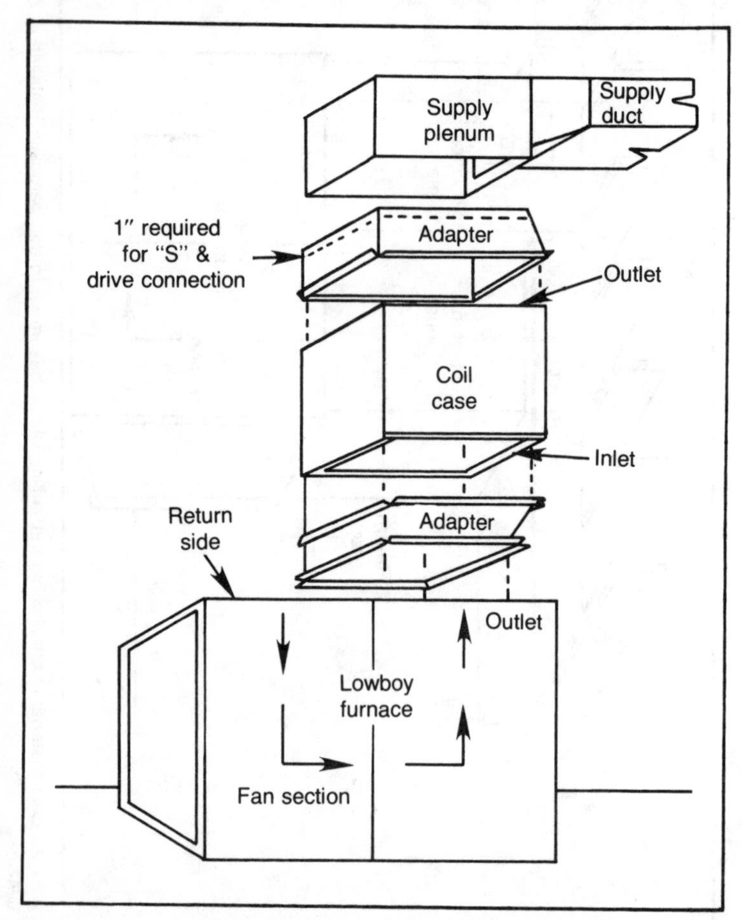

Fig. 3-55. Breaking a duct system into components will simplify measurements.

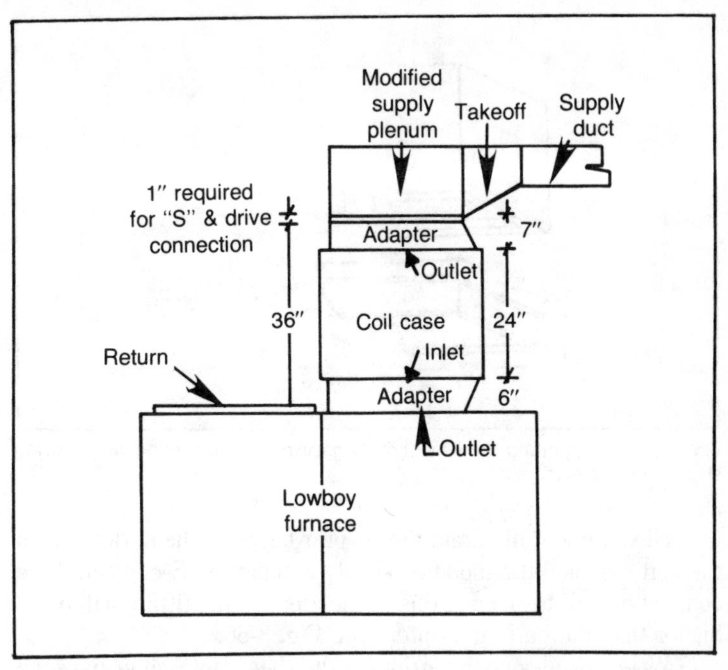

Fig. 3-56. Notice the takeoff shown in this diagram. The shape shown will reduce air friction losses to a minimum.

supply plenum out to accommodate the coil. Figure 3-54 should in some way resemble your particular furnace and duct system.

Refer to Fig. 3-54. The first measurement needed is the height of the space required for the coil and fitting installation. Leave at least 3″ between supply duct takeoff and the top of the space required. Let's say the dimension of the example is 36″.

Let's now assume that the height of the coil case we will be using is 24″. Also, 24″ subtracted from 36″ leaves us with 12″, which will be used up by the height of the two adapters. See Fig. 3-56. Note the 1″ space required for the S-clip and drive connection.

Now that we have the height of each fitting, let's decide what we will need for the opening sizes for each. We will start with the adapter between the furnace and the heat pump coil case.

Let's say the opening of the inlet side of the coil case is 22″ wide and 22″ deep, and the outlet of the furnace is 20″ wide and 20″ deep. Figure 3-57 illustrates exactly what the lower adapter will need to be.

Note the ½″ flange on both the top and bottom of the adapter. This is to be screwed to the furnace and coil case.

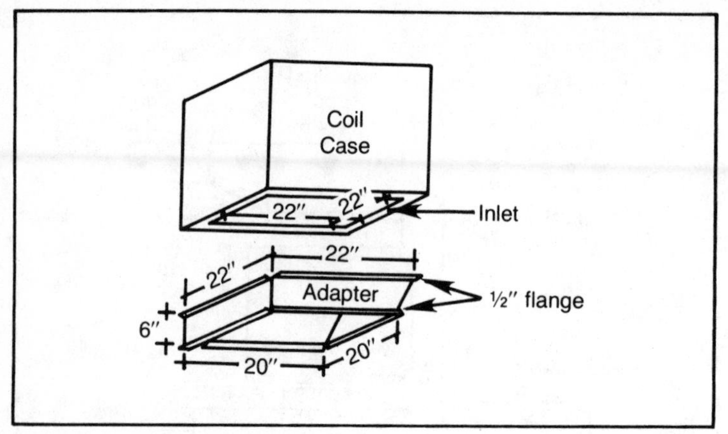

Fig. 3-57. All supporting sections of ducting will need to be connected by using sheet metal screws.

Next we will illustrate the adapter between the outlet side of the coil case and the modified supply plenum. An S-clip and drive connection will be used at this connection, so the fitting will be 1" higher than the bottom adapter. See Fig. 3-58.

When you pick up the fittings at the sheet metal shop, have the mechanic explain to you how the S-clip and drive connections are made. Have him turn the drive pockets on the top adapter.

Before the coil can be installed, it will be necessary for you to remove the portion of the supply plenum required. Be sure to leave the necessary 3" between takeoff and the top of the space required. For this project you will need two pairs of aviation snips—one that cuts to the left and one that cuts to the right.

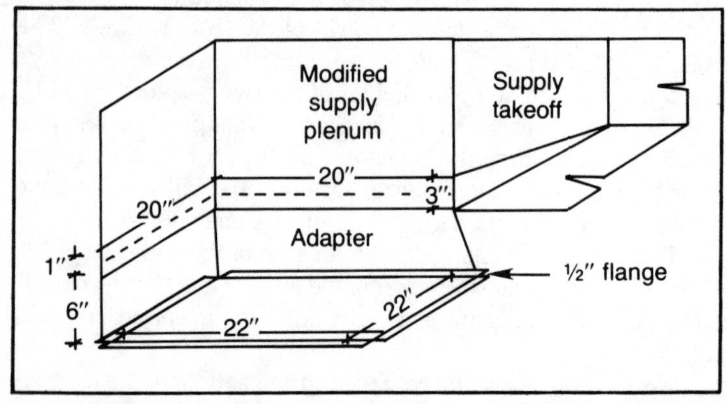

Fig. 3-58. Angled takeoff design is very important in proper air duct design.

After you have removed the necessary portion of the plenum, install the fittings and coil case. Turn the heat pump coil upside down and place the appropriate adapter on the inlet side of the coil case. Drill 1/8" holes through the flange into the case. Don't drill into the refrigeration lines. Insert #7 × 1/2" sheet metal screws into the drilled holes.

Place the coil and adapter onto the outlet side of the furnace. Repeat the fastening procedure described above.

The adapter on the top is more complicated to install because of the supply plenum above, but generally there is enough slack

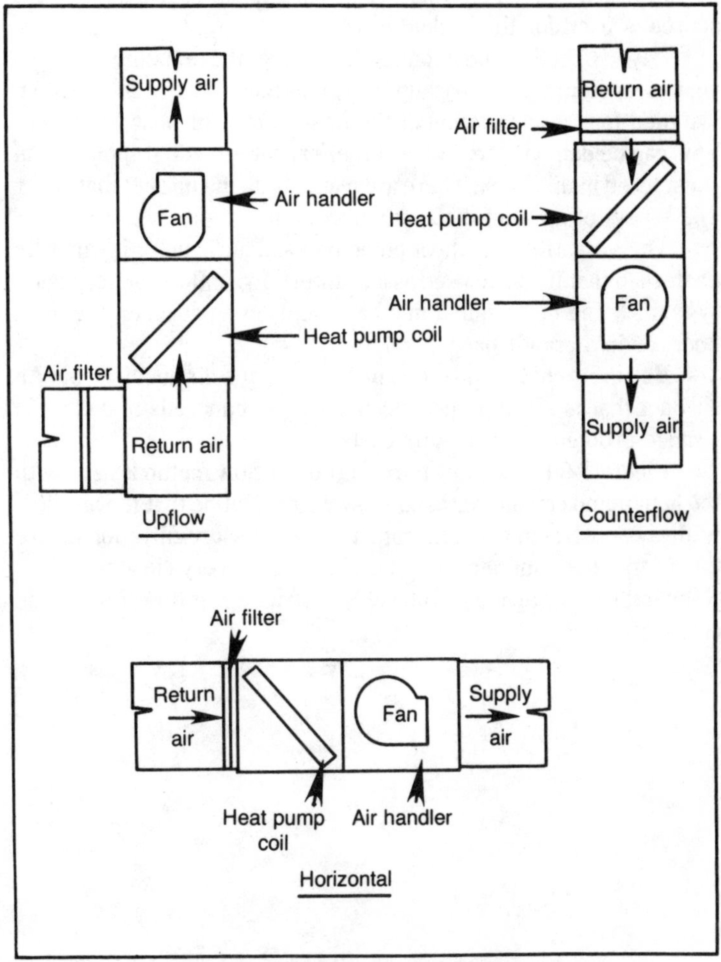

Fig. 3-59. Air handlers can be installed in several different configurations.

above the plenum to allow it to be forced up enough to get the adapter in place. You may need some help to hold the plenum up while you insert the adapter in place. When the adapter is in place, fasten it to the top of the coil case with screws. Using the S-clip and drive method, fasten the top connection.

VARIATIONS OF HEAT PUMP COILS AND AIR HANDLERS (NO BACKUP HEAT WITHIN THE SYSTEM)

This system is used if you have electric baseboard heat or hot water heat where there is no air duct system in the home. The heat pump system is used for your primary heat, and your existing heat source is used for the backup heat.

If you install a heat pump and leave the existing ductless heating system for the backup heat, you must have a duct system installed for the circulation of the heat from your heat pump coil. This can be done by the do-it-yourselfer, but it is too complex to be considered in this book. There are manuals on the market that cover this aspect of air duct distribution systems.

The versatility of a heat pump coil and air handler is virtually endless in that it can be used as a counterflow, upflow, or horizontal system. It can be installed in a basement, in an attic, on the main floor, or in a crawl space.

Be sure to have a qualified person size the air duct for you. An air duct that is sized properly can save you hundreds of dollars in service problems and heating costs.

Figure 3-59 illustrates the different air flow methods used with the heat pump coil and air handler systems. Notice that the air filter is always upstream from the coil. This is to assure filtration before the air hits the confined coil fins. The fins are very close together. Keep lint from entering and possibly restricting air flow through the coil section.

Chapter 4

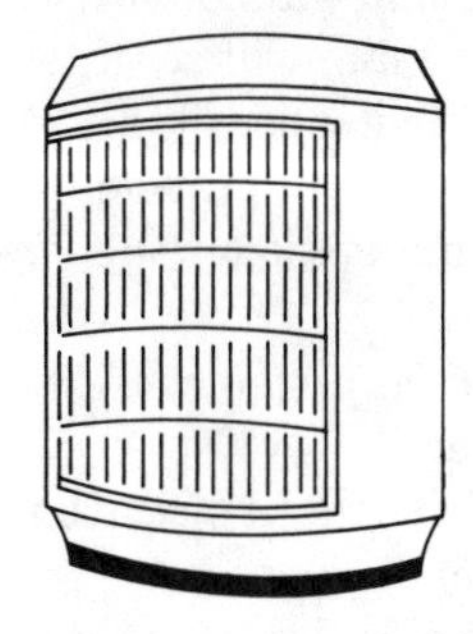

Mechanical and Electrical Installation

The outside section of the heat pump may be placed anywhere with in a 70′ area or radius of the indoor section. When the refrigeration lines are increased in length more than 75′, the friction inside the lines causes a loss of efficiency due to the added energy required to move the refrigerant through the lines or interconnecting piping. This problem can be corrected through engineering before the installation.

CHOOSING A LOCATION FOR THE OUTSIDE UNIT

The outside unit of the heat pump system moves several hundred cubic feet of air through it per minute. The air flow can not be diminished without greatly affecting the performance of the pump. The fans used in today's designs are normally propeller types because they offer quieter operation. A *squirrel-cage fan* (used in most furnaces) produces more air noise and requires added maintenance. It is probably advantageous to have a heat pump that discharges the air straight up from the outside unit. If the air is discharged vertically, most of the noise from the air movement will also rise with the air. The air flow of the unit is less likely to be restricted inadvertently. A minimum clearance must be maintained around the outdoor section for proper air flow. This minimum clearance is usually maintained at 1′ to 2′ from the air intake sections of the pump (refer to the manufacturer's notes). If the heat pump discharges the used air vertically, the air leaving the unit shouldn't strike any obstruction within 8′. If the discharged air does strike an object, it may be deflected back onto the heat pump, causing a loss of efficiency. Ideally the pump will not have any obstruction over it, and the nearest wall or shrub will be at least 3′ away.

Placing the outdoor unit is usually a compromise between what you can live with and what will provide the most savings. Listed below are some considerations in placing the outdoor unit (Figs 4-1 through 4-6).

- The unit should be placed as near as practical to the air handler or furnace.
- The unit shouldn't ever be placed under a bedroom window if you are a light sleeper.
- If you live in an area that has cold winters, don't place the pump where an icicle could fall into the fan.
- If the outdoor unit has to be partially under a roof eave, a short section of gutter material should be placed on the roof eave to divert water and ice. The water won't hurt the pump, but ice falling into it will.
- If the roof eave has a full section of gutter material, that long section may fill with ice. Install a heat tape to stop the icing if necessary.
- Don't put the heat pump in an area known to accumulate snow in drifts.
- If the average snowfall on the ground at any one time exceeds 2″, the pump should be raised on a snow curb. Snow curbs usually are from ½′ to 4′ tall, and they provide better air flow during snowing.
- If the outdoor section of the heat pump is placed a few feet from the home, it may be necessary to protect the refrigerant lines. Sometimes lines are in a place where people may step on them. If the lines are stepped on, refrigerant leaks may result that will require service.
- The unit should be in a location that allows sample service access to the unit.
- The heat pump will defost itself automatically during the winter. If the unit is under a window, it may fog it every time it defrosts. The defrost cycle will occur from once every 30 minutes to once a day, depending on the outdoor conditions and the brand of pump. If you like to look out the window, you shouldn't install the outside unit under a window.
- The outdoor unit shouldn't be put on a patio or near a

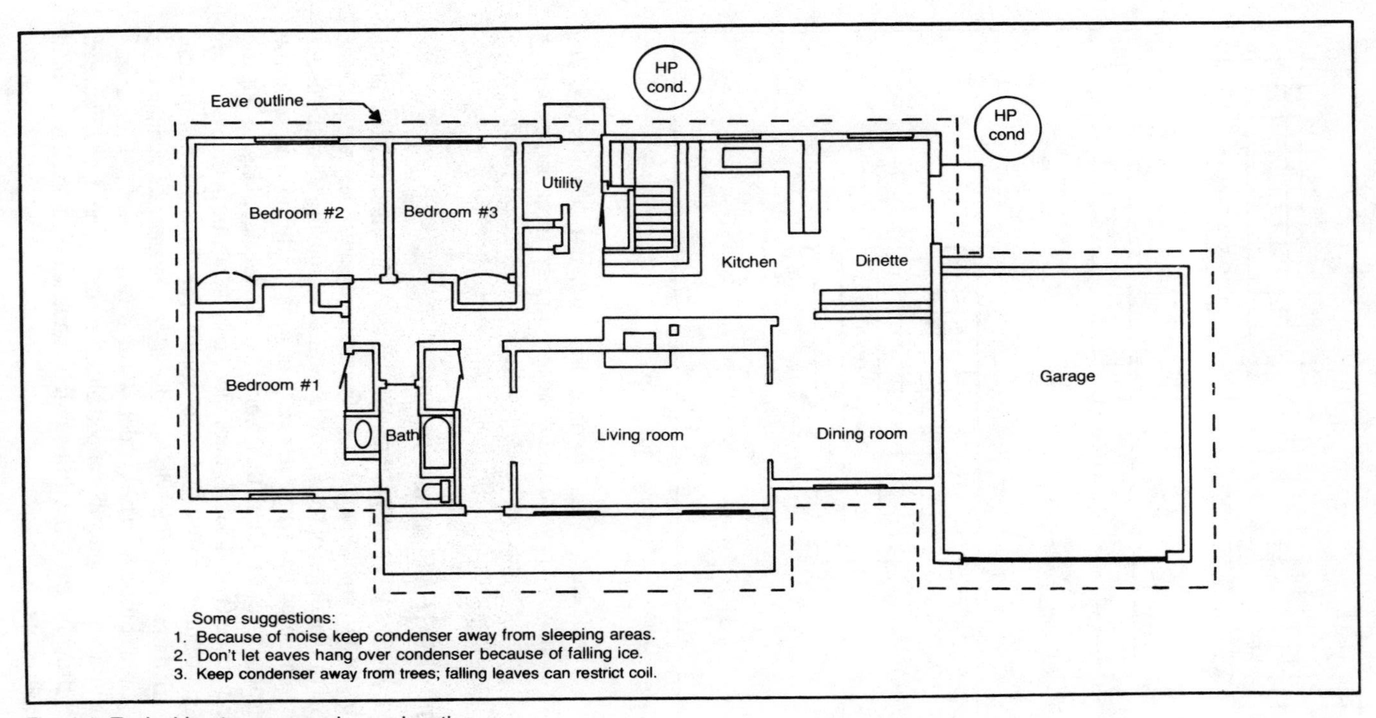

Fig. 4-1. Typical heat pump condenser location.

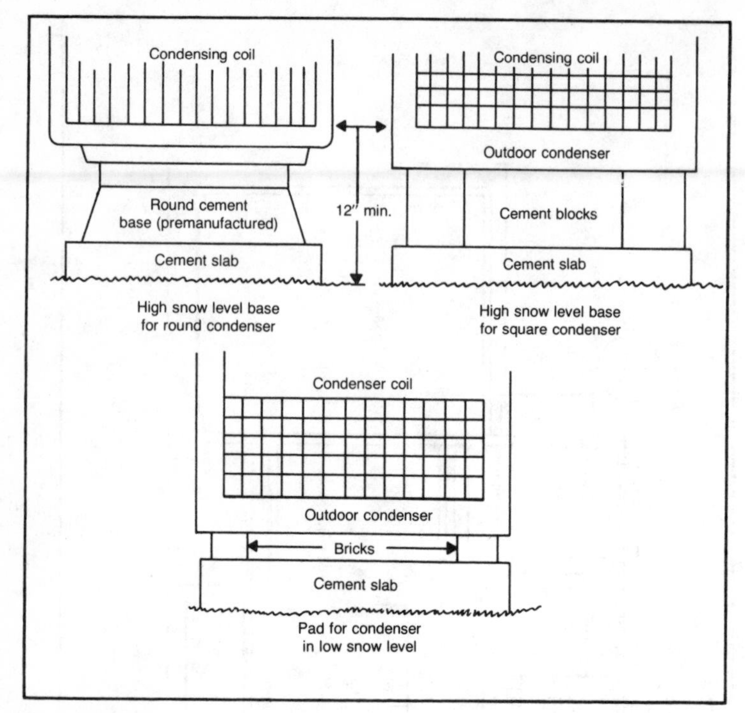

Fig. 4-2. Outdoor condenser pads.

sidewalk. When the unit defrosts itself, it melts ice. You may not have proper drainage in a warm area of the country. The water could refreeze in a colder area. Provide proper drainage.

CONDENSATE DRAINS AND PUMPS

Moisture is a by-product of the inside coil when the heat pump is in the air conditioning mode. The moisture must be drained away. Five to 10 gallons of water will be removed from the air in the house on an average day. If there is a floor drain in the area close to the furnace or air handler, that would be the logical place to put the condensate (Fig. 4-7). If your indoor coil is located in the return air duct, install a trap to stop the movement of air through the drain line.

If there isn't a floor drain in the area, a condensate pump will be required (Fig. 4-8). The condensate may be pumped to the outside to a utility sink and any trapped plumbing drain line. Do not exceed the lifting height capabilities of the condensate pump. Please read the directions.

PREMANUFACTURED REFRIGERATION TUBING

Presently there are two methods of connecting the outside unit to the furnace coil. Many refrigeration men will build their own piping system for each job. Another way is to install a ready-made set of refrigeration lines.

Building your own custom line set has the advantages of being inexpensive and attractive. The disadvantages are that the labor time is great, and the possibility of contamination is greatly increased.

If you choose to build your own line set, several precautions must be taken. This method is not recommended because of the hazard involved, plus related problems of contamination and pinhole leaks. Special solder is required, or the lines will blow apart. Further information can be obtained in books or schools that prepare a person for this part of the trade.

The ready-made line set can be installed by almost anyone, and it doesn't have the contamination or leak problems. The premanufactured line set also features reduced installation time. The tubing is shipped rolled in a coil. Take care in unrolling the tubing so that it doesn't kink. Follow the directions and take your time. Any

Fig. 4-3. A premanufactured heat pump snow pad is used to raise the pump above the level of average snowfall. Snow pads are also called snow bases.

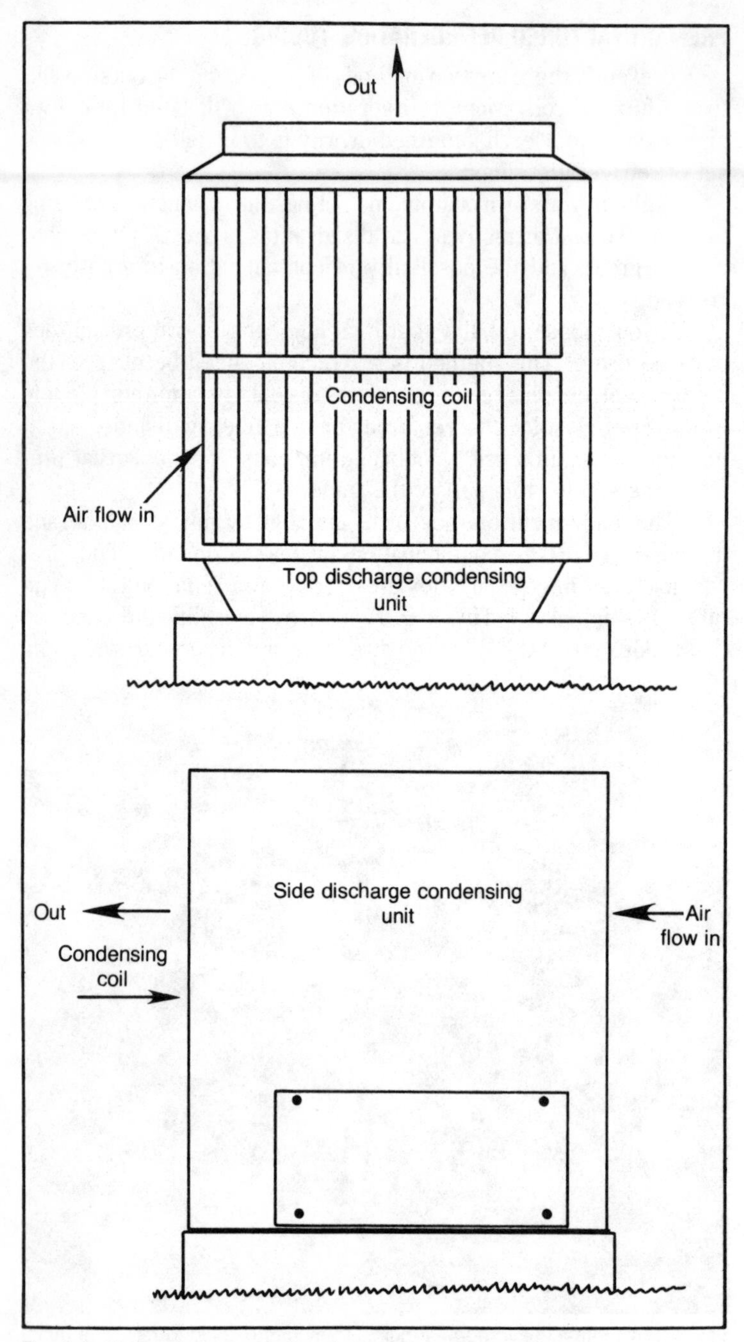

Fig. 4-4. Air flow in top and side discharge condensing units.

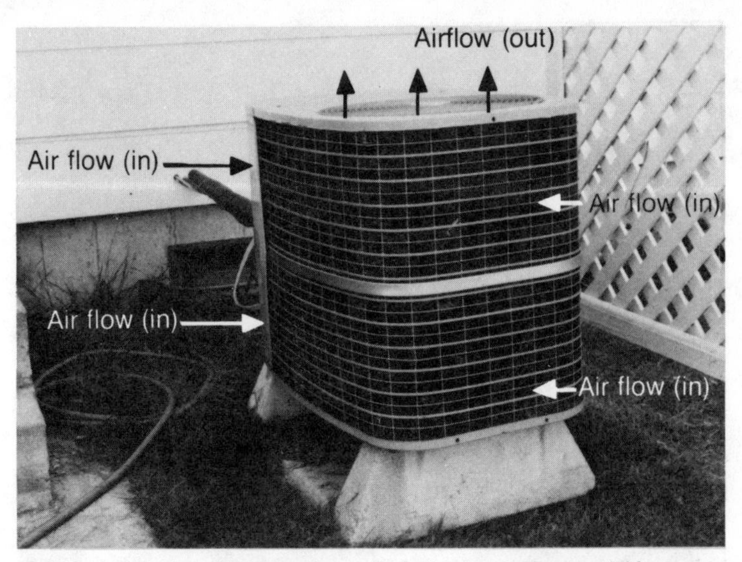

Fig. 4-5. Air flow through the outdoor unit of a heat pump is essential for proper operation. This unit processes the outside air in the directions marked by the arrows. Several hundred cubic feet of air are moving through this unit per minute of operation.

restriction such as a kink would cause reduced or total failure of the heat pump.

PRECHARGED TUBING CONNECTIONS

The indoor and outdoor units are to be connected together only with precharged interconnecting tubing. The units and the lines are

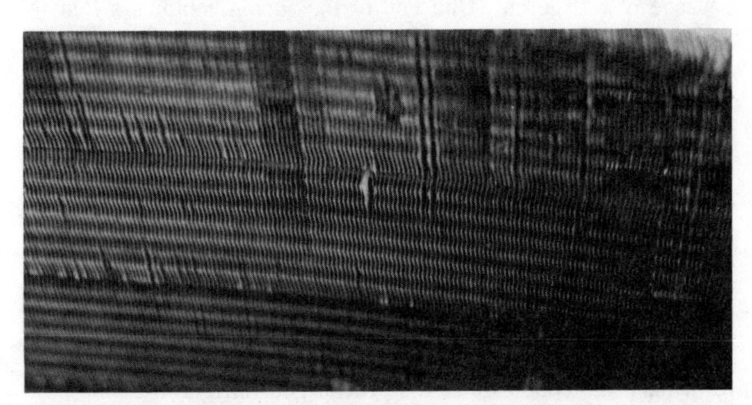

Fig. 4-6. Close-up view of the fin material of a coil in the interior of a heat pump. Fins increase the radiation area of a coil. Note the air passages between the fin material.

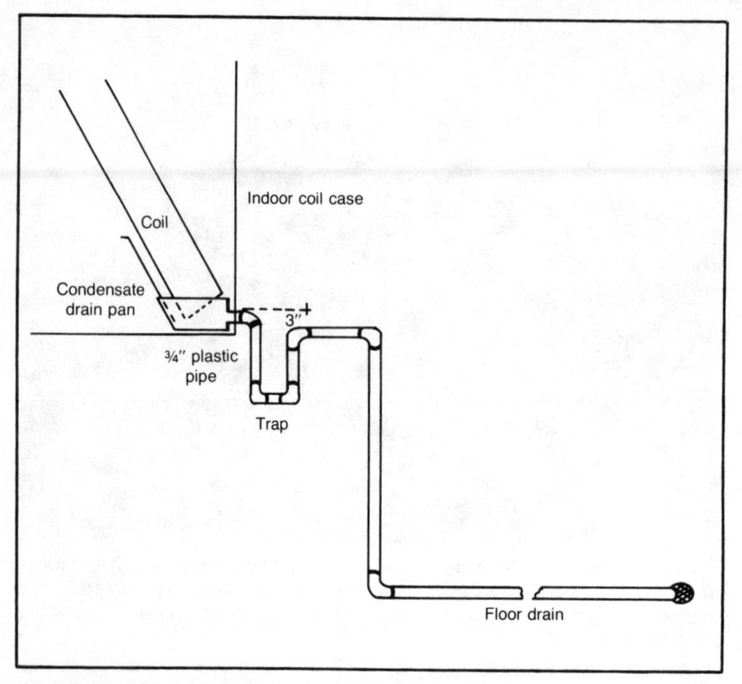

Fig. 4-7. Typical gravity condensate drain. Keep it free of debris so the condensate will flow into the drainpipe. If a floor drain is not available, install a condensate pump and drain to either the outside of the house or to a closed drain such as a drain vent or sink drain.

equipped with sealed one-time, quick-connect couplings which, when screwed together, will have their seals broken, allowing free passage of refrigerant. One end of the tubing contains a female fitting for connecting to the indoor unit. (Many manufacturers will use female fittings on both ends of the precharged tubing. The direction of refrigerant flow through the tubing is not important.) The lines are available in three different sizes. The length of the line varies from about 15′ to 60′, depending on the manufacturer. Refrigerant line in premanufactured sets is sold in 15′, 25′, 35′, 45′, and 55′ lengths. You need some idea of the distance between the outdoor section and the indoor section. Don't forget to include the distance to go around corners, etc.

Install the two refrigeration lines so that the air filter can be removed. Run the lines with as few bends as possible. Take care not to damage the couplings of the precharged refrigeration line set. Isolate the refrigerant lines to reduce noise transmission from the equipment to the building.

MATING QUICK-CONNECT COUPLINGS

- Do not proceed with this section of the installation unless you know and fully understand the instructions. It's possible to save installation cost by running the lines to the indoor and outdoor units, leaving the actual connection of the lines to the start-up serviceman.
- Begin at either the indoor or the outdoor unit. Remove only one dust cap from a line coupling nut at a time. By removing one cap at a time, dirt and other contaminents are reduced within the system if an accident should occur (Fig. 4-9).
- Lubricate the coupling threads with a few drops of refrigerant oil before making the threaded connection. By applying clean refrigerant oil, a positive, nonbinding seal can be made easily (Fig. 4-10 and 4-11).
- Carefully screw the swivel part of the tubing coupling onto the male half. Make sure the threads are not crossed. The swivel can be rotated one or two turns by hand before the seal in the coupling begins to open. Once the seal is

Fig. 4-8. Condensate pumps come in many sizes and shapes. This unit has a large water-holding capacity. The water is held in the black plastic box under the labeled pump motor. Note the clear plastic line connected to the pump; water leaving the pump travels through this line. When the white polyvinyl chloride (PVC) drain lines were originally installed, the condensate pipe was run across the floor drain. Later the customer decided to add a condensate pump to the system. The water trap, shown under the condensate pump, was needed for the original floor drain installation. The PVC riser (a vertically traveling pipe) causes the trap under the pump to be ineffective. This particularly pump installation serves two indoor coils, and the pump is held to the metal by sheet metal screws. Condensate pumps must be leveled for dependable operation.

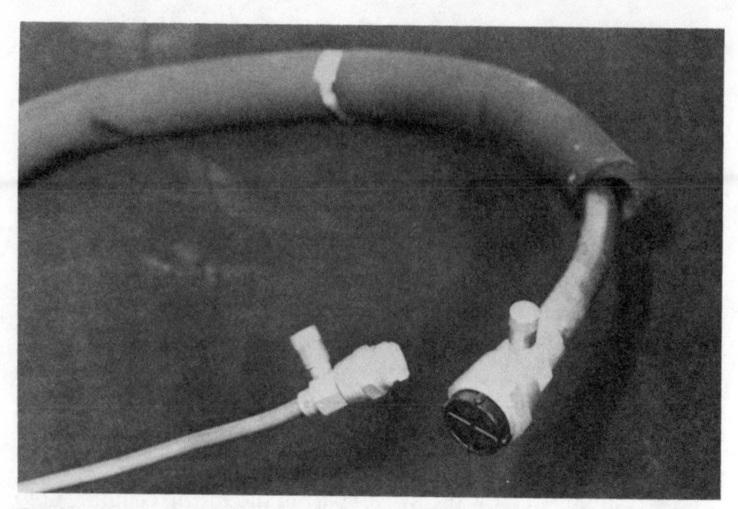

Fig. 4-9. When removing the plastic duct caps, be careful that you don't puncture the inner seal of the tubing under the cap. Precharged line sets contain refrigerant under pressure within each line.

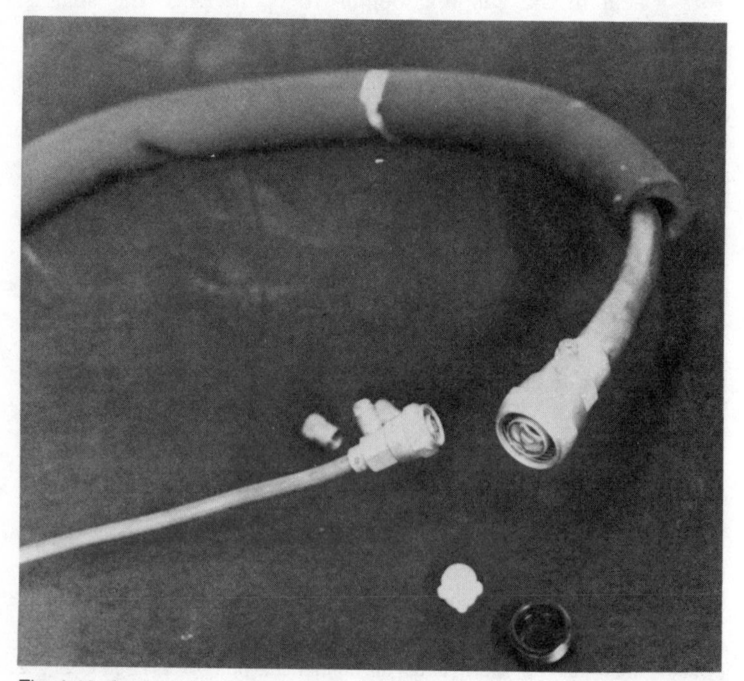

Fig. 4-10. Apply one or two drops of oil to the male section of the line coupling. Note the schrader caps. The schrader capped end of the line set is normally attached to the outdoor unit.

Fig. 4-11. The outdoor section of a heat pump has the following connections that must be made for its control and energy transfer: the vapor line connection, the liquid line connection, the thermostat cable entering the outdoor unit, the electrical power cable feeding the outdoor section of the pump. Notice the rubber covering on the vapor line.

punctured, the lines must be mated. If not, the charge (refrigerant) will be lost (Fig. 4-12).

- Continue to tighten the connection until the swivel begins to become snug—about four additional turns. To be sure the tubing doesn't turn, hold a wrench on the hex nut of the tubing coupling (Fig. 4-13).
- Tighten the coupling approximately one-fourth to one-half turn past the point of a snug fit. Do not overtighten; damage may result.
- Repeat this procedure with the other three couplings in the system. Remove the dust caps just before making the connections.

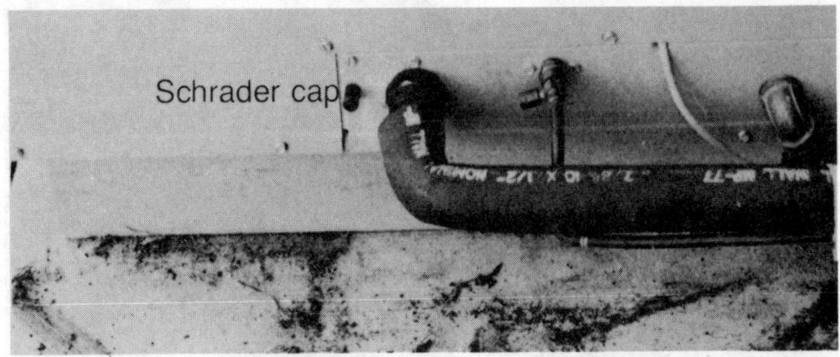

Fig. 4-12. Make sure that the line set mechanically aligns with the heat pump connections. Make large radius bends in the piping unless you own a tubing bender.

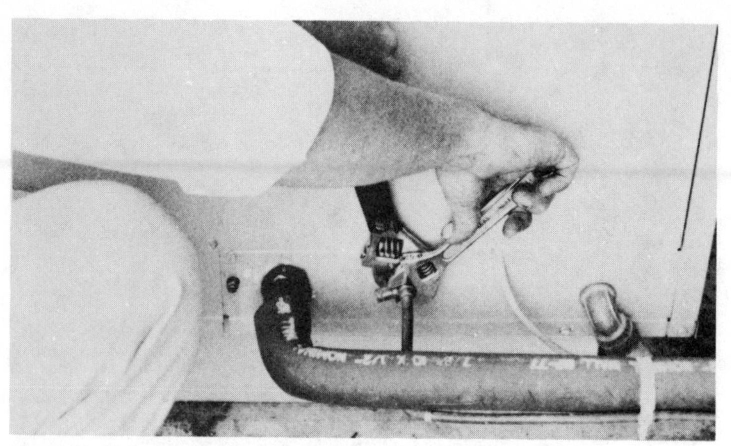

Fig. 4-13. Using two wrenches to tighten the connection.

- Test each coupling for leaks after connecting. A soap and water solution is good for testing leaks if there isn't a chance that the solution will enter any refrigerant passages. If the system contains pressure and the fittings are mechanically tight, the solution won't enter the interior of the piping, but it will bubble even if a small leak exists (Fig. 4-14).
- Check the flare caps on the schrader fittings of the interconnecting lines. The caps should be fingertight. Overtightening will cause the cap seal to fail (Figs. 4-15 and 4-16).

Fig. 4-14. When premanufactured tubing sets are used, check the mechanical connections for leakage. Use a soap solution when the system is under pressure from the refrigerant gas.

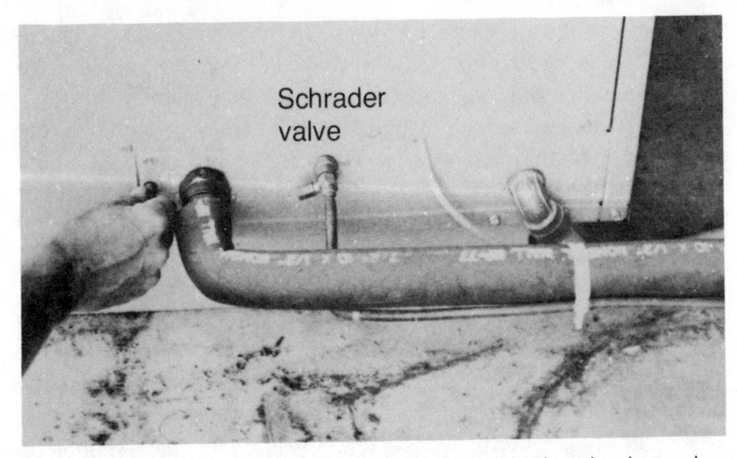

Fig. 4-15. Schrader caps should only be hand tightened unless they have a hex head on the cap.

SECURING THE REFRIGERANT PIPE

When you have roughed in the refrigeration pipe and it is in the area where you want it, secure the line permanently. During the rough-in, the lines are tied into place with a scrap piece of wire. Then the U-clamps can be installed with minimal adjustment problems. Also, the liquid and vapor lines can be strapped or taped to each other. Nylon wire ties give the best result if they aren't overtightened. Overtightening will cause the nylon strap to cut through the foam rubber covering on the vapor line. If a rip or cut

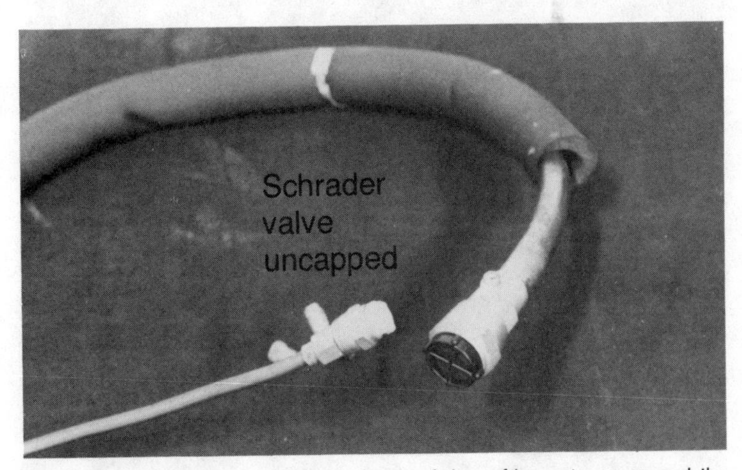

Fig. 4-16. Overtightening causes leakage of the refrigerant gas around the gasket material of the cap.

develops within the vapor line foam, the defect should be taped, glued, or covered to prevent air from entering the fracture. We recommend that the large line (the vapor line) be installed first; it's much harder to work with than the small liquid line. Don't kink the lines during installation, or heat pump failure will occur. If a kink should develop, the line must be repaired to allow full flow of the refrigerant.

The vapor line is covered with a black foam rubber insulation installed at the tubing factory. The insulation retains heat during the heating mode and stops condensate from forming on the pipe during the cooling function (Fig. 4-17).

When the liquid line and vapor lines are secured together (it's possible to include the control cable with the line set), determine where the U-clamps will be installed. The lines should be supported every 6′ to 8′ of the piping horizontal run (Fig. 4-18). The U-clamps should be a minimum of 1″ wide, so that they won't cut through the vapor line insulation. The best results are attained when both pipes are wrapped with a scrap piece of insulation. This provides isolation of the lines from the clamp. The noise from the pipe will be kept out of the floor joists (Fig. 4-19).

If the control cable is run with the refrigeration lines, be sure that the cable is not between the clamp and the lines. We have found that with time, the insulation will break down under moderate pressure from the pipe, thus shorting out the control cable.

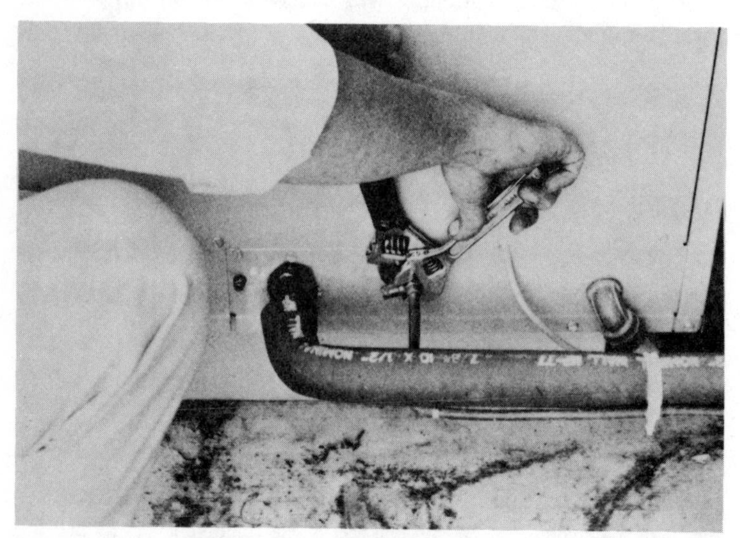

Fig. 4-17. The vapor line is covered with foam rubber insulation.

Fig. 4-18. Lines are strapped to the ceiling.

INSTALLING THE THERMOSTAT

Normally the thermostat will be mounted where the old thermostat was mounted—on an inside wall, 4′ to 5′ above the floor. If you have a finished wall on which you want to mount the thermostat, you will have to *fish* the wall. You have to have an unfinished ceiling in the space below where you want to mount the thermostat. Determine where the studs are by tapping on the wall. Drill a ¾″ hole through the Sheetrock between the studs. Drill a pilot hole next to the wall. Use a small-diameter drill that is long enough to penetrate the floor. When the pilot hole is made, leave the drill bit in place. Go downstairs, locate the drill bit sticking through the floor, and measure over to the approximate center of the wall. When you

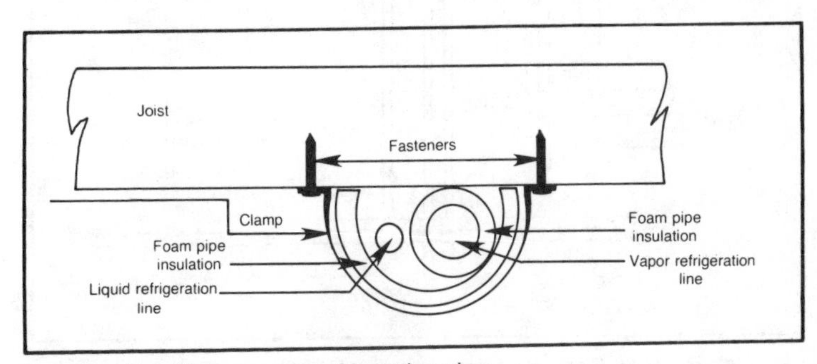

Fig. 4-19. Detail for securing refrigeration pipe.

have found the center of the wall, drill a ¾″ hole from the basement, through the wall plate, and into the wall.

Find a small chain about 8′ long and feed about 7′ of it through the hole in the Sheetrock upstairs. Secure the chain so that it can not be pulled through the Sheetrock hole. Go downstairs with a flashlight and a straightened coat hanger. Bend a small hook on one end of the hanger and insert it into the hole through the floor. By working the coat hanger around throughout the inside of the wall, you should be able to hook the chain and pull it through the lower ¾″ hole. Twist the thermostat wire around the chain and tape it. Go upstairs and pull the wire up the inside of the wall with the chain. It's okay to run a thermostat wire in a return air duct, but never in a supply duct. See Figs 4-20 through 4-22.

TOOL LIST

- ¼ or ⅜ electric drill.
- Extension cord.
- Tape rule.

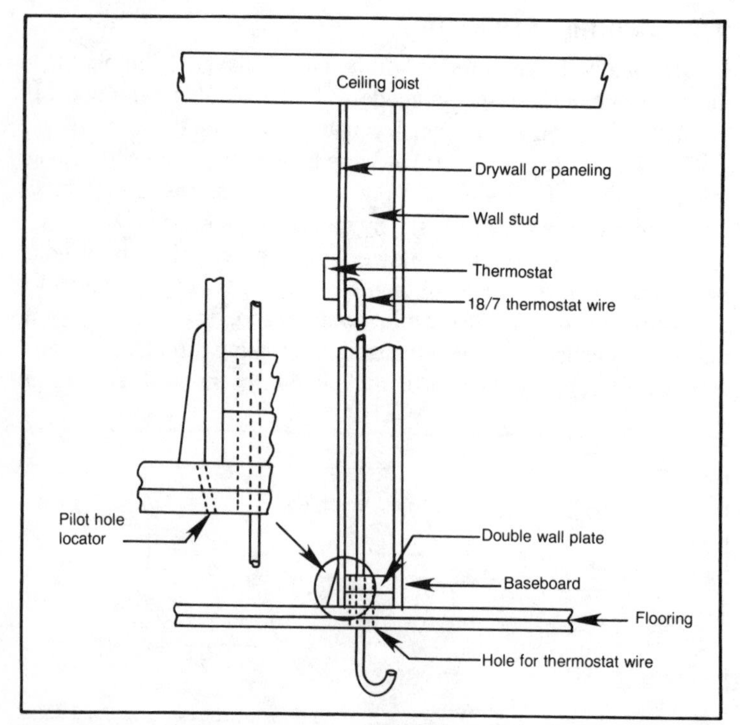

Fig. 4-20. Detail for thermostat and wire installation.

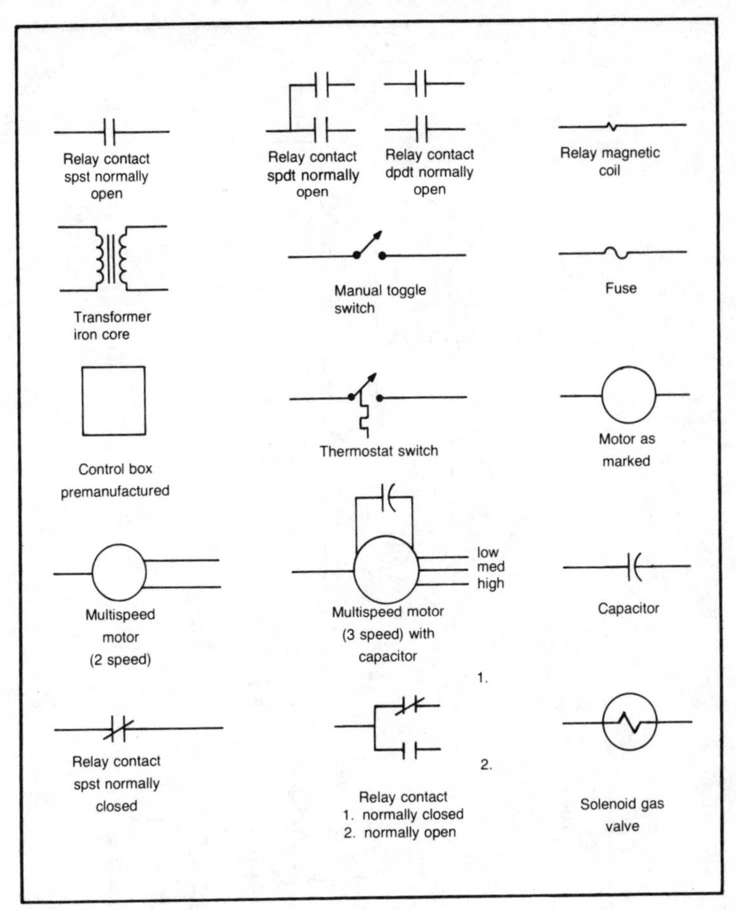

Fig. 4-21. Electrical symbols.

- Level.
- Hammer.
- Tin snips—straight and aviation.
- Screwdriver—Phillips and standard.
- Straightedge/yardstick.
- Scratch awl.
- Marking pen.
- Drill bits—wood and metal.
- Crescent wrench—12″.
- Water pump pliers—12″.
- Voltohmmeter.
- Allen wrench set.

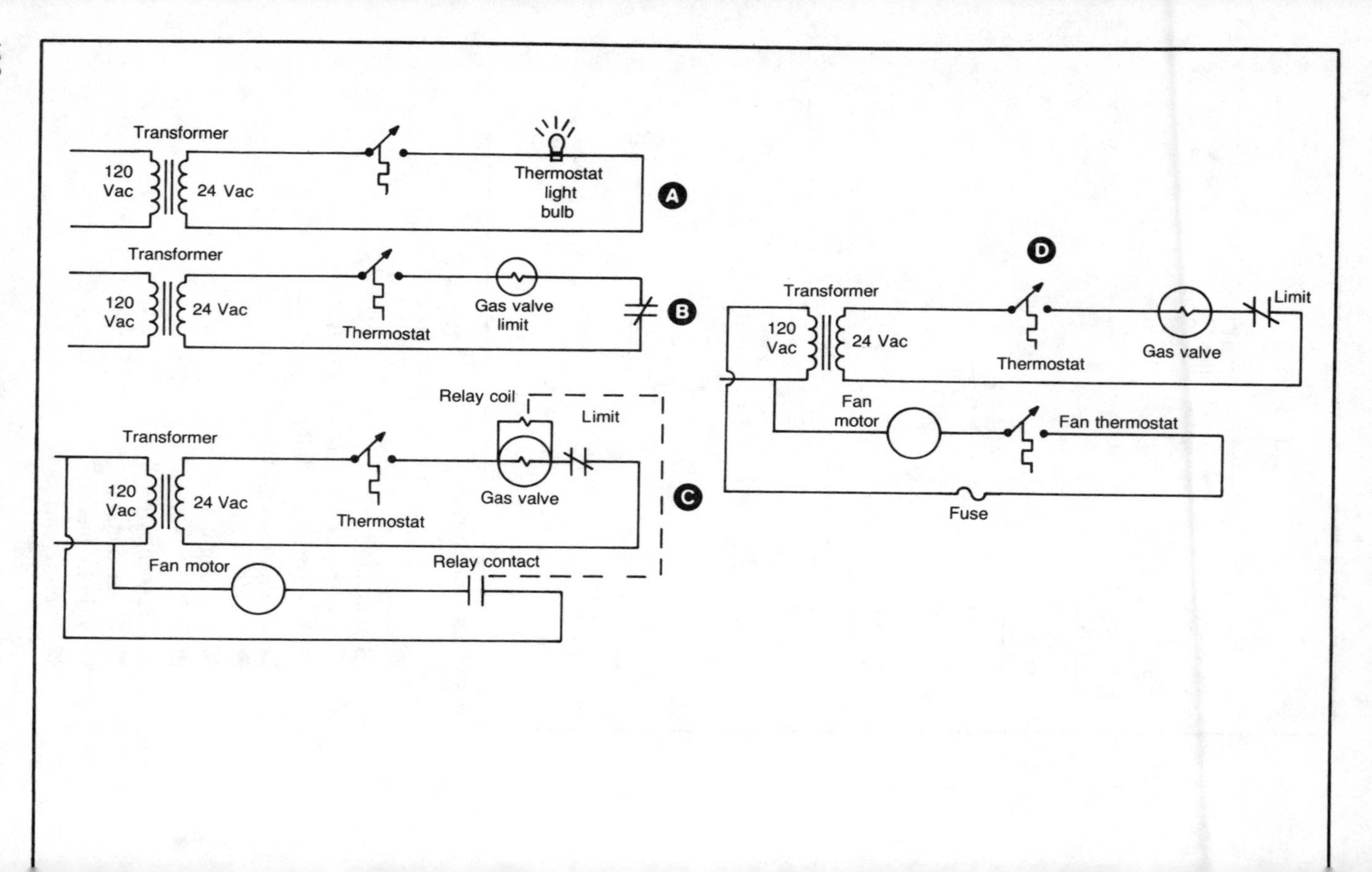
Transformer
120 Vac
24 Vac
Thermostat light bulb
A
Transformer
120 Vac
24 Vac
Thermostat
Gas valve limit
B
Relay coil
Limit
Transformer
120 Vac
24 Vac
Thermostat
Gas valve
C
Fan motor
Relay contact
D
Transformer
120 Vac
24 Vac
Thermostat
Gas valve
Limit
Fan motor
Fan thermostat
Fuse

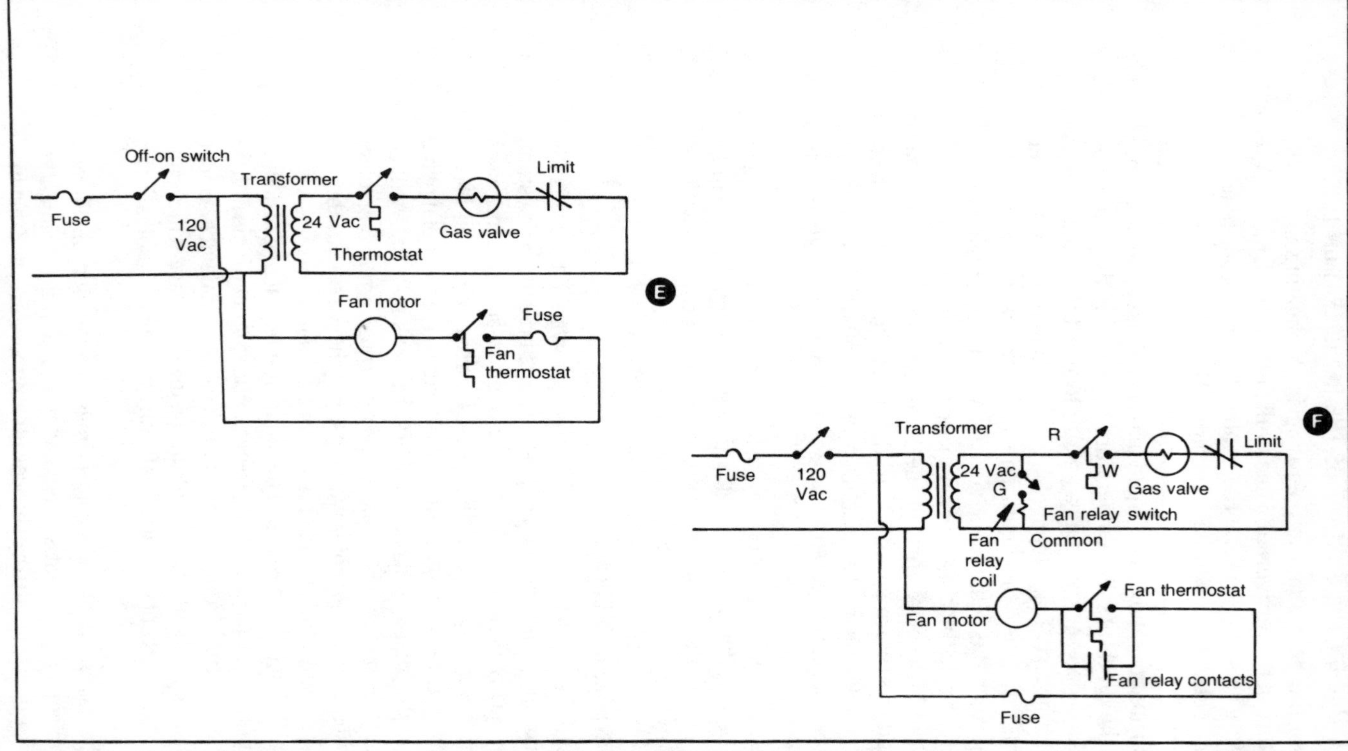

Fig. 4-22. Heating control wiring diagrams. (A) Thermostat operating a 24 Vac light bulb. (B) Gas furnace circuit without a fan. (C) Gas furnace circuit with fan motor that starts with the gas flow. (D) Gas furnace circuit that operates from the temperature increasing inside the furnace. (E) Gas furnace circuit with fuse in the main power supply and fan circuit. (F) Gas furnace circuit with a fan relay activated by the low-voltage summer fan switch circuit.

TEMPERATURE ON THE SURFACE OF THE REFRIGERATION LINES

When the heat pump is operating in the heating mode, the temperature of the refrigeration lines will fluctuate with the outdoor temperature to some degree. The cooler the outdoor air temperature, the cooler the lines will become. The vapor or large line generally will run from 175° F. to about 90° F., depending on the outdoor air temperature. The liquid line should never be more than warm to the touch. As the temperature increases in the liquid line, the pressure increases, causing a greater load on the compressor. A dirty air filter, dirty fan, slipping fan belt, or an improperly adjusted fan speed will cause the liquid line to be too warm. Excessive temperature on the liquid lines will cause compressor failure if the condition persists. Increase the fan speed if the liquid line is warmer than 90° F. Make sure that all the registers are open, clean air filters, and the return air supply to the furnace is free of obstructions. Depending on the manufacturer, 280 to 450 CFM of air flow is required per ton of refrigeration. The air flow has to be right for your installation.

When the heat pump is operating in the cooling mode, the vapor line will be cold but not covered with ice. The liquid line will be slightly warm. If the vapor line or the indoor coil is covered with ice, the air flow is too low or the unit is low on Freon.

AIR, MOISTURE, AND DIRT

One of the primary causes of refrigerant system failure is the presence of air, moisture, and dirt. When air is present in a system, moisture usually is, too. The presence of air in any system raises the discharge pressure. Moisture causes corrosion and oil breakdown. Dirt and other foreign matter cause excessive wear of moving surfaces in the high-speed compressors. Ferrous materials will be attracted by the motor. These materials may cause grounding (electrical) or shunting of the motor windings. For these reasons there is no substitute for a clean, dry, refrigerant system.

Some servicemen have used the undesirable practice of adding alcohol or other antifreeze agents to hermetic systems. This is done to prevent moisture from freezing in the expansion device. Under no circumstances should such solutions be used; damage will result. If any antifreeze agents are used, all warranties are automatically void.

In the factories every effort is made to remove all air and moisture from a refrigerant system and to prevent dirt from entering the system. Eliminate the sources for air, moisture, and dirt first rather than later when making refrigerant circuits in the field. You should understand dehydration, evacuation, leak testing, and charging requirements.

If you are in doubt about a certain section of the installation, call in a professional and pay him by the hour. Also, do not start the unit yourself. Talk to the people that sold the unit to you and arrange the steps you must follow to secure warranty protection.

REFRIGERATION LAWS

- There is no such thing as cold until the thermometer registers—460° F. All temperatures above this point are considered to contain heat. We use the term cold every day, but it would be more accurate to say that we have less heat.
- Heat always flows from the warmest object in an area to objects of lesser warmth.
- Any time a liquid changes to a gas or vapor, it must release the heat that it contains. The heat is carried away in the vapor. Whenever a liquid changes its state from a liquid to a vapor, heat is pulled out of the liquid that remains. This law applies to all liquids that change states. When your skin is wet and a breeze vaporizes that moisture, a cooling effect is felt. The moisture has changed to a vapor, and cooling has occurred. The changing state principle is the basis for all operation of refrigeration equipment.

HEAT PUMP REFRIGERANT FLOW DIAGRAM

See Fig. 4-23. If you follow one set of arrows or the other, you will find the path of the refrigerant through the various components in the heating or cooling mode. All heat pumps are not exactly the same because small parts differences exist, but the flow diagram is basically the same for all heat pumps.

You really don't have to know this information to install a heat pump. We felt that you might want to see how the flow changes from heating to cooling.

Reversing Valve. The reversing valve has one function—to reverse the refrigerant flow direction so that heating or cooling may be selected. The reversing valve is the heart of the heat pump system (Fig. 4-23).

A reversing valve can be energized either during the heating or the cooling mode, but not both. Some pumps will provide switching during the cooling cycle, yet others will switch during the heating cycle. By switching, we mean that the reversing valve is energized

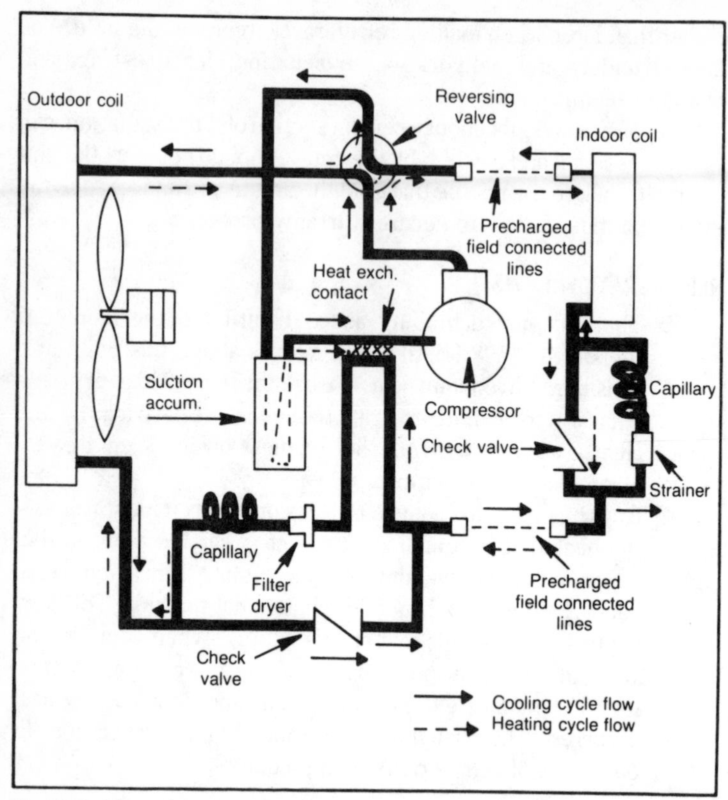

Fig. 4-23. The refrigerant flow pattern in one of the better heat pumps on the market.

electrically. Brand Y may apply power to the valve during heating cycles only, but the valve would not be energized during cooling and defrost. If we energized the valve during cooling and defrost, it wouldn't be energized during the heating mode. The valve is only powered during one cycle or the other.

As far as the valve is concerned, it really doesn't make any difference if it's energized or not. It is thought to be better, though, if the valve is energized during the heating mode (Fig. 4-24).

Strainer. A strainer is a metal screen used to catch particles that might plug the capillary tube.

Check Valves. A check valve is a device that only allows the refrigerant to flow in one direction through the valve. If the refrigerant is flowing opposite to an open check valve, the valve will be forced shut. This will stop the flow and force the refrigerant to take another circuit. Most check valves are usually the ball check type.

Capillary Tube. The capillary tube is a section of copper tubing of a selected length with a known diameter hole through the tube. By selecting the hole diameter and the length of the tube, you can select the amount of refrigerant that flows through that tube. A capillary tube is a refrigerant metering device. A thermostatic expansion valve has the same function as a capillary tube, but the metering is more accurate.

Suction Accumulator. When refrigerant vapor is returning to the compressor, it must not contain any liquid refrigerant. The function of the accumulator is to catch any liquid that might get back to the compressor and hold that liquid until it turns to a vapor. If liquid reached the compressor intake, valve damage could result.

OUTDOOR THERMOSTATS

The purpose of an outdoor thermostat is to turn the furnace off when the outdoor temperature is above a certain level. This type of thermostat is similar to the one inside your home. It is simply a switch activated by temperature.

When your home is sized in relation to the heat pump, a heat usage rate is found. Normally the heat pump will warm the home,

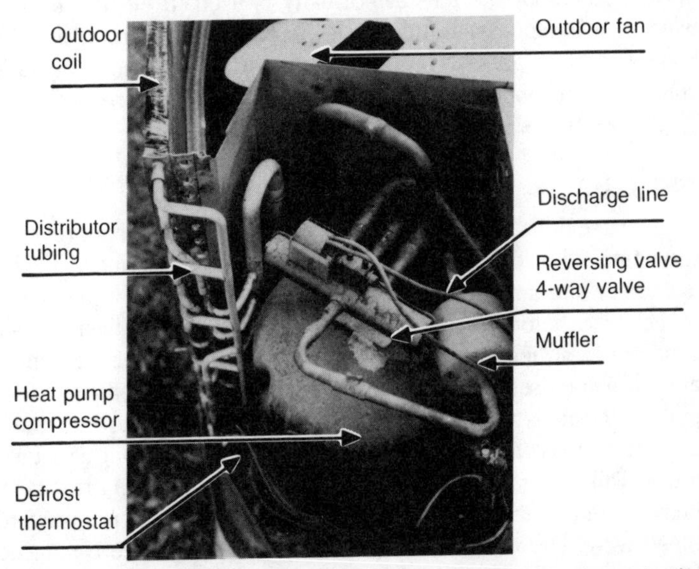

Fig. 4-24. The outdoor unit of a heap pump will contain the following parts: the outdoor coil, the outdoor fan blade, the reversing valve with its electrical solenoid, the heat pump compressor, the muffler to quiet the heat pump compressor, the distributor tubing for the outside coil, the defrost thermostat, and the discharge line of the compressor.

with an outside temperature of 30° F. (plus or minus 4°), without the aid of the furnace. Below the 30° F. outside temperature, the heat pump will require assistance from the furnace to maintain the temperature inside the home.

The advantage of an outdoor thermostat is simply to reduce operational costs. Why use backup heat when the heat pump has the capacity to do the job at a greatly reduced cost? If a pump is to give you the best economy possible, a comfortable temperature should be found, then the thermostat setting shouldn't be changed. It won't damage the equipment to vary the thermostat setting, but it's more economical to leave that setting at one point. If the indoor thermostat setting is changed, without the addition of an outdoor thermostat, the furnace and the heat pump could operate together (assuming that the indoor coil is in the return duct).

As gas, oil, and electricity are more expensive per unit, it's not recommended that you operate the heat pump and furnace together. The outdoor thermostat will lock out the furnace or backup heat above a certain outdoor temperature. As the outdoor temperature lowers, the outdoor thermostat switches, allowing the furnace to supplement the pump as needed. If oil or gas is used for backup heat, only one outdoor thermostat normally is used. If electric heat is used as a backup, each individual element in the furnace can be staged on at progressively lower temperatures. Features of stage control are economy, extended parts life, and the reduction of electrical surges into an electrical furnace.

TIME DELAY

A time delay is a mechanical or electrical device that prevents the pump or air conditioner from turning on for a set period. The advantage of a time delay is to allow internal refrigerant pressures to equalize before starting the pump compressor. When an air conditioner or heat pump operates, very high pressures are generated. If a compressor is running, then turned off and on again, the piston of the compressor has a tremendous pressure to start against. Unless enough time has passed before attempting a restart, a fuse will probably blow because the compressor will draw very high current. Time is an important factor for the high pressure to bleed away. The addition of a time delay will stop pressure-induced fuse blowing. The time delay will also reduce mechanical wear because of reduced stress. If a heat pump unloads (releases the high pressure), a time delay is of little use. A time delay isn't usually included with the pump.

Chapter 5

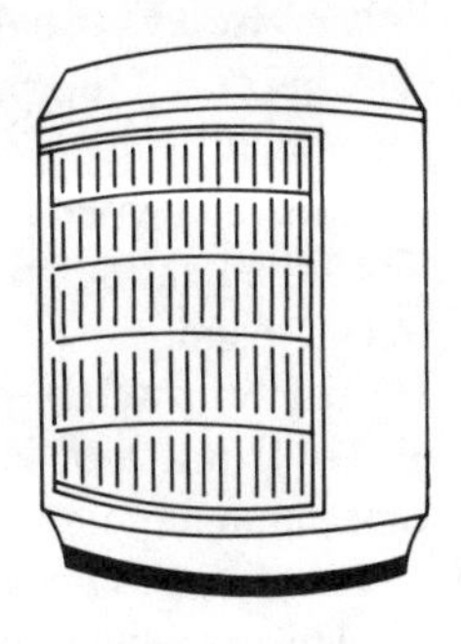

Electrical Control Function

Many types of control systems are used to operate the furnace/heat pump combination. The control system will affect the performance of your system. If installed incorrectly, it will greatly reduce the efficiency. The following information is general and is only meant to give you a basic background in control function.

CONTROL SYSTEMS

In the Pacific Northwest the winters are severe; as a result, 100 percent backup heat is required. If the heat pump fails or if the weather is below −10° F., an alternate heat source must be available. In the southern states a backup heating system may not be required because of the mild winter temperatures. In many of our illustrations the duct systems are connected to furnaces, but an air handler could be in the place of the furnace if a large amount of backup heat isn't required. (An air handler is a fan inside a case without its own heat source.) Backup heat can be supplied in any percentage of the required output of the heating system. The amount or existence of backup heating is dependent on the average winter temperature of a geographical area. Specific heating and cooling requirements can be obtained from your heating dealer.

Basic Control Function: A Heat Pump with an Air Handler, Indoor Coil in the Return Duct, and No Backup Heat. The thermostat turns the heat pump on or off depending on the temperature setting. This action occurs in the heating or cooling mode. Operation of the indoor fan is always included with the heat pump.

Thermostat and Fan Switch in the Automatic Setting. A manual switch is usually located on the thermostat to control the fan manually. The indoor fan will turn on and off with the outdoor section of the heat pump. If the fan switch is set in the on position, the fan will run continually. The heat pump will function independently and on thermostatic demand.

Basic Function: Heat Pump with an Electric Furnace, the Coil in the Return Air Duct, and Heating Mode. In this type of installation we want minimum electric furnace heating element operation. As in other control systems for heat pumps, a heat pump thermostat is used. This particular type of thermostat has been designed for furnace (air handler)/heat pump combinations. When a signal for heat is sent by the thermostat, the heat pump and furnace fan are turned on, starting the heating cycle. If the room temperature declines still further, the furnace will operate with the heat pump, supplying the needed additional Btu.

To achieve this staged operation, a temperature difference is built into the thermostat. If the thermostat is set at 70° F., at about (plus or minus 1°) 69° F. the heat pump and fan will begin to function in the heating mode. If the indoor temperature continues to fall 2° to 3°, the second stage of heating will be turned on to supply the additional heat needed in the living area. Two sensing elements are used in the thermostat to permit a staggered turn-on sequence. In stage one the heat pump and fan operate at the same time. In stage two the heat pump, fan, and furnace become operational. The heat pump will always run before the furnace and after the furnace has satisfied its portion of the heating load. Return mounting of the coil is ideal so far as economy is concerned. The maximum benefit is drawn from the heat pump in this configuration.

The Cooling Cycle. When the thermostat calls for cooling, the heat pump switches into the cooling mode—mechanically and electrically. The heat pump begins to function with the indoor fan. The pump will continue to run until the thermostat setting has been satisfied. The heating function must never operate at the same time. The heating function is the stronger of the two modes in terms of Btu. The furnace or air handler forced air fan may provide better results if it runs continually.

Basic Control Function: Gas or Oil Furnace, with the Heat Pump Coil in the Return Air Duct, and Heating Mode. In this type of installation we want minimum gas or oil consumption.

As in other control systems for heat pumps, a heat pump thermostat is applied as the automatic function switch. When a signal for heat is sent by the thermostat, the heat pump and furnace forced air fan are turned on, starting the heating cycle. If the room temperature declines past the thermostat temperature setting by 2° to 3°, the furnace heating system will be activated. If the heat pump is able to maintain the temperature within 1° or 2° of the setting, only the pump and furnace fan will operate to maintain the best economy level. By having a two-stage heating operation wired into the control circuit, we only use backup heat when absolutely necessary. Again, two sensors are used in the thermostat to give staggered operation of the two heat sources. In stage one the heat pump and forced air fan operate at the same time. In stage two the heat pump, forced air fan, and furnace operate at the same time. The heat pump should always run before the furnace and after the furnace has satisfied the signal of the second stage for heat, if required to maintain space temperature. Return mounting of the coil is ideal as far as economy is concerned. You get maximum benefit from the heat pump in this configuration, unless rusting is a problem, due to summertime humidity levels. Consult your heating dealer to see if return coil mounting is possible in your area.

Basic Control Function: Gas, Oil, or Electric Furnace, with a Heat Pump, Using Outdoor Thermostats. An outdoor thermostat reduces heating costs by locking out the backup heat source. When the heat pump load (heating and cooling) is known, then the outside air temperature that the heat pump can handle and effectively heat the living space is determined. That heat is maintained at a constant level. The lower or lowest temperature at which the heat pump can maintain the living space temperature is called the balance point. In the colder climates of the country, backup heating operation above the balance point temperature isn't desirable for reasons of economy. How do we keep the furnace turned off if someone should raise the temperature setting on the thermostat? The answer is to use an outdoor thermostat. The outdoor thermostat will open the control wire that energizes the secondary heating source above a manually set temperature on the outdoor thermostat. Balance points usually fall in the range of 28° F. to 35° F. for the lower heating temperature of the heat pump in most residential applications. Installation of outdoor thermostats is covered in the heat pump manufacturer's wiring instructions. Gas and oil furnaces require one outdoor thermostat for operational control. Electric furnaces can be wired with one outdoor thermostat for each heating element or several stages of lockout.

Basic Control Function: Gas or Oil, with a Heat Pump Coil in the Supply Duct. Because the coil in this example is located in the supply air duct, some interesting changes occur. It won't be possible for the heat pump and the furnace to operate at the same time. The heat generated by the furnace is intolerable in the refrigerant circuit of the heat pump. The coil is downstream from the furnace regarding air flow. An intolerable condition is created because the refrigerant absorbs heat from the furnace, and there is a great increase of refrigerant pressure. This pressure increase is far beyond the working pressure limits of the heat pump system. Because the heat pump and furnace can not operate together, the control system must not allow both units to operate together at any time. The furnace or the heat pump can work separately, but never together.

The heat pump will switch on as the room cools. If more heat is lost from the room, the heat pump will switch off and the furnace will turn on, with the forced air fan running all the time. When the room approaches a normal temperature, the furnace will turn off and the heat pump will operate until the need for heat is satisfied. In stage one the heat pump and the forced air fan will operate. In stage two the furnace and the forced air fan will operate. This stage function is built into the thermostat, but additional wiring is required for proper functioning in the either/or mode of operation.

Because the refrigerant in the heat pump circuit is sensitive to temperature, the warmth from the furnace heat exchanger will be felt during every cycle where the furnace operates and then stages back to heat pump operation. The warmth from the furnace will cause an increase in the operating pressure of the heat pump until the furnace heat exchanger cools. Increased operating pressure means increased compressor mechanical wear, plus heavier than normal electrical power requirements.

If a time delay was used to stop the heat pump from turning on for a predetermined time, we could extend the life span of the compressor. The amount of time that the heat pump should remain off, with the furnace fan still running to cool the heat exchanger, generally will be five to eight minutes. The composition of the heat exchanger will decide how long the pump should remain off in waiting for the furnace to cool. The heavier the metal in the heat exchanger, the longer it takes to cool the furnace.

To determine if a time delay will stop heat pump operation for the required time, temperatures must be taken in the supply section of the ductwork. If the furnace air flow is of the proper volume for heat pump operation (see Chapter 2), the test is as follows. Start the furnace and allow it to blow heated air for 15 to 20 minutes. During the time that the furnace is producing heat, insert a thermometer into the supply duct, 5′ to 10′ from the furnace, toward the outlet side of the system.

After waiting for the recommended time, turn the thermostat down to stop the fire from operating. The furnace fan should continue to blow warm air from the furnace, even though the fire has gone out. Note the time that the fire was shut down and also the temperature reading on the thermometer. If the temperature falls to 100° F. or less in under eight minutes, a time delay will be sufficient to protect the heat pump. If the temperature doesn't fall quickly enough to meet the eight-minute mark, make sure the air filter is clean and the registers are open to allow maximum air flow. If you find that the system is free of air flow restrictions, but the temperature is still excessive after eight minutes, then a control other than a time delay is necessary to keep the heat pump turned off until the temperature is reduced. Installation of a time delay into a heat pump control circuit takes five minutes or less.

The control wire that carries the signal to tell the compressor to operate is almost always the terminal marked Y, on the heat pump low voltage wiring block and the terminal marked Y within the thermostat. The Y-terminal rule applies to 98% of all residential heat pump manufacturers; Y operates the switch that turns on the compressor. To install a time delay of the two-terminal type, cut the wire interconnecting the Y terminal in the thermostat to the Y terminal within the heat pump. Each end of the cut Y wire will be attached to the terminals of the two-wire delay device. Only use time delays in low-voltage wiring. Observe the time delay instructions included with the control. Three- and four-terminal time delays are on the market, but they are more difficult to install. If you want to keep the compressor turned off, electrically open the Y control wire.

Another method of controlling compressor operation is to install a temperature-sensitive switch in the Y wire. A limit control is a switch that opens when temperature limits are reached. A limit also has a sensor mounted within the supply duct, so accurate air temperature measurements can be observed by the control. The limit switch is a two-wire device, and it may be installed just as the time delay by cutting the Y control lead and attaching each end of the cut wire to each terminal of the limit switch. The temperature that

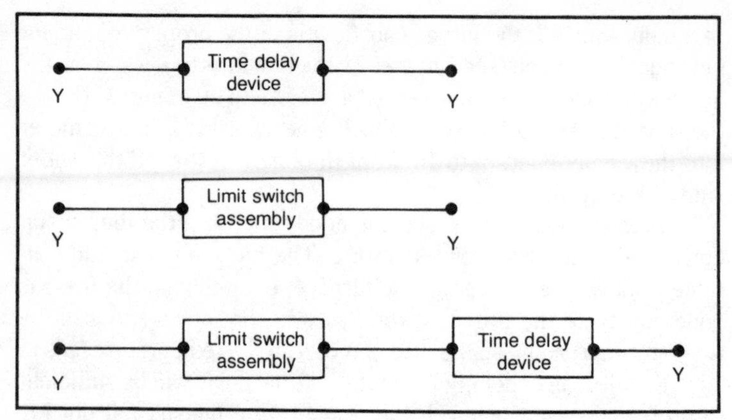

Fig. 5-1. Time and temperature control of a heat pump compressor. Special purpose switches are used to prevent early turn-on of a heat pump with the coil in the supply ducting.

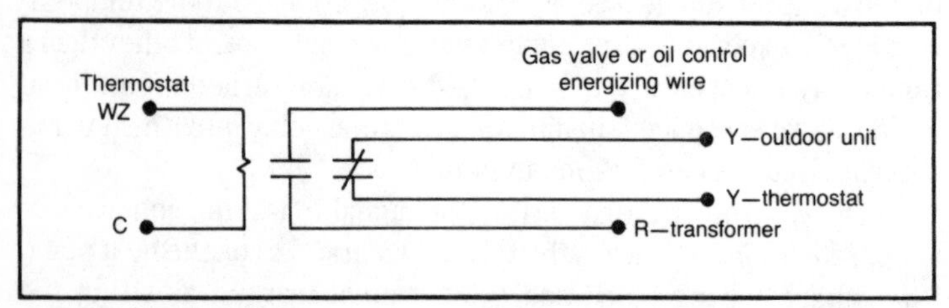

Fig. 5-2. This wiring circuit only allows the backup heat or the heat pump to operate. It does not allow both units to function simultaneously. This circuit is only used when the coil is in the supply ducting. This control circuit is for furnace only or heat pump only.

the limit opens the Y circuit should be set at 90° to 100° F. When the duct temperature falls below the limit setting, the switch will make (close). The compressor circuit will be energized.

A time delay is preferred as a control. If high temperature is a problem, only the limit switch will cure the effect. It's possible to connect the limit and time delay in series in the Y lead, but more expensive. See Figs. 5-1 and 5-2.

STAGED OPERATION OF HEAT PUMPS WITH VARIOUS HEATING SYSTEMS

When a heat pump operates, the amount of heat delivered to the interior of the home is dependent on the temperature of the outdoor air. If operation of the heat pump is above the balance point of the particular installation, the heat pump can carry the entire heating load of the living space. When heat is called for by the thermostat, the heat pump will turn on and supply that heat as needed. When the thermostat is satisfied, the heat pump will switch off and wait for the next cycle. Second-stage heating won't be needed to augment the output of the heat pump.

If the same pump should operate when the outdoor temperature is below the balance point, the heat pump will require assistance in maintaining the living space temperature. Balance point temperatures generally fall in the area of 28° to 35° F.

Control systems are rather straightforward in design when the heat pump coil is located in the return air ductwork. As heat is needed in the living space, the thermostat will turn on additional heat as it's needed, either first or first and second stages depending on demand.

When the heat pump coil is located in the supply ductwork, the control needed is more detailed in scope. The heat pump and the secondary heat source can't operate together because of the properties of the refrigerant. The heat from the furnace will blow across the heat pump indoor coil, thus adding temperature to that coil and refrigerant. When temperature is added to the refrigerant, it will increase the pressure within the refrigeration piping. The additional pressure is highly undesirable. In this type of system we operate the heat pump or the furnace, but never at the same time. Following is an explanation of basic control function.

Heating Sequence of a Heat Pump with a Low Outdoor Temperature Indoor Coil in the Return Duct—Gas, Oil, or Electric Backup Heat

- The living space temperature is at 70° F. and falling slowly. The thermostat is set at 70° F. within the living space.
- The indoor temperature falls to 69° F. The furnace fan turns on, and the outdoor section of the heat pump also starts to function in the heating mode.
- Warm air begins entering the living space through the duct system from the indoor heat pump coil.
- The living space continues to lose heat. The temperature is now at 67° F.
- The second stage of heating is activated at a differential of 3° from the main thermostat setting. Second-stage heating is turned on. The heat pump and the furnace fan continue to run and deliver two stages of heat (stage one, the heat pump; stage two, the backup heat or furnace).
- A gain of temperature is noted within the living space. The indoor temperature is now at 69° F. with the addition of

second-stage heat. Second stage turns off because of the reduced differential—less than 3°.

- The heat pump and the furnace fan continue to operate, with heat flowing from the ductwork into the living space.
- The living space is again beginning to lose heat in excess of the heat pump output into the living space. The heat pump can't carry the load by itself. The second stage turns on once again, bringing the space temperature bace to 69° F.
- The outside temperature is measured at 21° F. Ice forming on the outside coil of the heat pump.
- The outside unit of the heat pump senses the buildup of ice on the surface of the outdoor coil. A whooshing sound comes from within the outdoor section. The heat pump is in the defrost mode. The furnace fan and the compressor continue to run. The outdoor fan stops while the unit is in defrost.
- The backup heat begins to function during the defrost cycle to temper (warm) the air coming from the indoor coil. (Since the defrost function is air conditioning, very cold air could come from the supply ductwork during defrost if the backup heat source isn't turned on with every defrost cycle.)
- The ice and frost are melting from the outside unit coil. The defrost cycle continues until a liquid temperature of 80° F. is attained, leaving the outdoor coil.
- Another whooshing sound is noted coming from the outdoor section of the heat pump; the second stage heat turns off. The outdoor fan again begins to function. The liquid temperature has reached the predetermined 80° F. level, and the defrost cycle has been terminated. (Defrost cycles are terminated by time or temperature.)
- The living space temperature is falling slowly; it's measured at 68° F.
- At a differential of 3° from the thermostat setting, the second-stage heating function will provide additional heating capacity.

This sequence will repeat until the outdoor temperature exceeds the balance point of the heat pump. The defrost cycle within the heat pump may occur at any point in its running mode during heating. This step-by-step listing has been prepared to assist you in visualizing normal, indoor coil in the return duct operation.

Heating Sequence of a Heat Pump with a Low Outdoor Temperature Indoor Coil in the Supply Duct—Gas or Oil Backup Heat

- The living space temperature is at 70°F. and falling slowly. The thermostat is set at 70°F. within the living space.
- The indoor temperature falls to 69°F. The furnace fan turns on, and the outdoor section of the heat pump also starts to function in the heating mode.
- Warm air begins entering the living space through the duct system from the indoor heat pump coil.
- The living space continues to lose heat; the temperature is now at 67° F.
- The second stage of heating is activated at a differential of 3° from the main thermostat setting. Second-stage heating is turned on, and at the same instant the heat pump turns off, delivering second-stage heat only. (With the coil in the supply, the heat pump can't operate while the furnace is functioning.)
- There is a temperature gain within the living space. The indoor temperaure is now at 69° F. with the addition of second-stage heat. (The backup heat source will have much greater output than a heat pump in low operating temperatures.) The second-stage heat is turned off because the thermostat senses less than a 3° differential from its setting selected by the homeowner. Note that the temperature within the living space is still lower than the thermostat selected temperature by 1°. The heating system in this case will revert to first-stage heating, which is the heat pump. As mentioned earlier, if a coil is placed in the supply duct, it's important to install a time delay in the Y lead to the heat pump. The time delay will allow the furnace to cool before the heat pump operates. The time delay will allow the furnace to cool before the heat pump operation starts, when the heating system switches from second to first-stage heating.
- The furnace fan operates continuously through first, second, and then back to first-stage control functions. When the heating system switches from second to first-stage operation, since our examples has a time delay, an eight-minute waiting period will occur before the heat pump again functions. During the eight-minute wait, the furnace will have time to dissipate the heat that it's holding within the heat exchanger.

- After the timing period, the heat pump compressor starts and produces heat. The outdoor temperature is measured at 21° F. in this example.
- As the outdoor section of the heat pump operates and the indoor temperature again begins to fall slowly, the pump is operating below its balance point.
- While the pump is operating in the heating mode, it senses that it has a buildup of ice and frost on the outside coil. The heat pump turns the outdoor section fan off, the four-way valve switches with a whooshing sound, and the second-stage heat starts to buffer the air leaving the indoor coil. (When a heat pump defrosts, it actually becomes an air conditioner for a short period. If the second-stage heat isn't turned on with the defrost function, the air leaving the supply ductwork will be very cold.)
- The ice and frost are melting from the outside unit coil. The defrost cycle continues until a liquid temperature of 80°F. is attained, leaving the outdoor coil.
- Another whooshing sound is coming from the outdoor section of the heat pump. The second-stage heat turns off. The outdoor fan again begins to function. The liquid temperature has reached the predetermined 80°F. level, and the defrost cycle has been terminated. (Defrost cycles are terminated by time or temperature.)
- After the defrost cycle, the heat pump reverts to first-stage heating, thus providing heat to the living space.
- The living space temperature is falling slowly; it's measured at 68° F.
- At a differential of 3° from the thermostat setting, provide additional heating capacity.

This sequence of events will repeat until the outdoor temperature exceeds the balance point of the heat pump. The defrost cycle within the heat pump may occur at any point in its running mode during first-stage heating. This step-by-step listing has been prepared to assist you in visualizing normal, indoor coil in the supply duct operation.

LOW-VOLTAGE CONTROL TRANSFORMERS

A *transformer* is an electrical device that transforms an input voltage and current into a lower or higher output voltage and current rating. A transformer uses electromagnetic induction to effect the voltage and current transformation from one circuit to another.

Step-down transformers are used in heating and air conditioning work. The secondary or output voltage is of lesser value than the input or primary voltage. You usually will find a 24-volt transformer within heating equipment (across the output leads). If a unit contains electronic control circuits, a special transformer may be required. In almost all cases the transformer voltages will be marked on the body of the transformer.

The input voltage found within heating equipment for transformers can be one of the following: 120 Vac (volts—alternating current), 208 Vac, or 240 Vac in residential equipment.

The output voltage of typical heating and air conditioning transformers is 24 Vac. If measured without a load, it's not unusual to find up to 28 Vac across the secondary wire leads. The higher than rated secondary voltage won't affect the function of the circuits for control.

A transformer is rated in VA units (volt-amperes). If a transformer is 24-Vac output at 10 VA, the output current rating is 0.417 amp. If a transformer is rated at 24 Vac, 40 VA, its current output rating is 1.67 amps. Note that VA/Vac = current rating of the secondary. Use only 40-VA transformers for the heat pump applications. Lesser transformers may not have enough power to handle the control circuits.

THERMOSTATIC CONTROL

Thermostatic control of various heating and cooling functions is nothing more than single-pole, multiple-throw switching. The control voltage, the R side of the transformer, goes to the R terminal within the thermostat. The R terminal is switched to W for heat, to Y for cooling, to G for constant fan operation, and to O plus Y for heating in some heat pumps. In heat pumps the "O" terminal and the Y terminal are connected to R for heating only, or for cooling only, but not both. The four-way valve piping determines the heating or cooling function when the valve is turned on by the O terminal. In most cases the four-way valve will be energized in the cooling mode or in the defrost function. The O terminal within the thermostat, when connected to the R terminal, completes the circuit of the four-way or reversing valve.

The common side of the low-voltage transformer is used within a thermostat only for the emergency heat light or, in special cases, for a common connection for electronics within the thermostat. The common side, C or B, is never the leg of the transformer that is automatically switched to provide control function. Refer to the wiring diagrams within your furnace to determine which wire of the transformer is connected to the common side of the control circuit.

Several circuit diagrams have been developed to show the switching action of thermostats on the market. These diagrams are simplified and meant for instructional purposes only—to show how the R terminal is switched to the other terminals within the thermostat.

LOW-VOLTAGE WIRING TERMINALS AND HEAT PUMPS

Standardized low-voltage wiring unfortunately isn't a reality in the heating and air conditioning industry. Even though the function of the control wires is the same, manufacturers use their own terminal marking systems.

The thermostat, furnace or air handler, and the outdoor section of the heat pump have a need to be interconnected for proper control function. The low-voltage wiring that interconnects the various sections of the heat pump can be considered a nerve system. Each of the wires controls an electrical device that has a planned purpose.

The main low-voltage control wires have the following functions in all heat pumps. Remember that the markings between manufacturers are different. The circuit diagrams will not look the same, but the functions of the various wires are similar. If in doubt, the circuit diagram should be studied to determine the control function of each wire.

- One of the transformer output leads is a common. This wire is common to all devices that use the control voltage for activation. The common side transformer terminals may have C, LC2, or B marked next to the terminal.
- The switched side of the transformer (24-volt or other control voltage) is usually connected to the R terminal or possibly the LC1 wiring point (see the Thermostatic Control section).
- Activation of the heating function is normally carried by a wire terminal marked W, X2, or LC5. More than one stage of heat may be carried by the circuit. Another terminal may be used as an additional heating control.
- The Y terminal is used predominantly to start the heat pump compressor in the heating or cooling mode. If a time delay or a temperature switch is used to keep the heat pump turned off, the Y wire should be opened by the control.

More than one stage of cooling may be included with your heat pump; it may be marked Y2. Staged heating units are now on the market.

- The reversing or four-way valve—the valve that switches and causes the heat pump to operate in heating or cooling modes—will have one control wire. This wire usually is marked O.
- During the defrost function, the indoor heat will be activated to offset the air conditioning duct temperature. A terminal is included in the outdoor unit to turn this function on and off.
- When the heat pump is put into emergency operation, a control signal is sometimes fed to the outdoor section of the heat pump for deactivation of that unit. This switch function may be taken care of within the thermostat; it depends on the manufacturer's design.
- The voltage for the emergency heat light usually is fed to the thermostat from the outdoor section to the thermostat. Again, one-wire will be used.
- The G terminal is normally a control terminal used for control of the furnace or air handler forced air fan circuit.

Because of the nonstandardized wire terminals in the heating industry, the wiring variations are endless. A complete book could be filled with just wiring diagrams to show each situation that might be needed in a particular application. After you purchase a heat pump, study the wiring diagrams and the manufacturer's recommended interconnection of the various components. Ask the dealer to supply you with a low-voltage wiring diagram for the type of unit that you're buying if you plan to do the control wiring yourself. All wiring must conform to the National Electrical Code and local codes.

If you want to stay away from the control wiring, a substantial savings can still be achieved if you just run the control wire from the thermostat to the furnace, then from the furnace to the outdoor section of the heat pump. Usually eight wire thermostat cable is used (eight wires of 18-gauge wire in a cable) for the control functions. About five percent of the units in the field will use 10 wire cable. When a technician comes to your home to start the heat pump, it's simple to connect the cables that you have run to reduce your labor investment.

The heat pump, thermostat, and most additional parts will have installation instructions. Study them along with the material in this

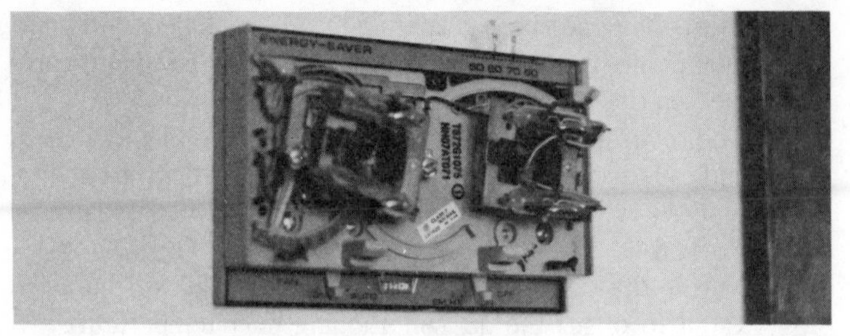

Fig. 5-3. Heat pump thermostats are manufactured in several different configurations to match the control requirements of the various heat pumps.

book and visualize the finished product. Refer to the installation checklist, then it's just one step at a time.

HEAT PUMP THERMOSTATS

Heat pump thermostats are multistage, with manual or automatic function switching (Figs. 5-3 and 5-4). In a manual switchover model, the heating or cooling function of the thermostat has to be manually selected. The automatic version will turn the heating or cooling on as needed. Most heat pump thermostats include a fan switch that determines if the forced air fan (indoor) operates automatically with the heat pump or on continuously. Another feature is an emergency heat switch that turns the heat pump off and allows furnace operation only. Depending on the configuration of the four-way valve piping, a heat pump thermostat will energize the four-way valve coil (solenoid) in either the heating or cooling mode. Seventy-five percent of all heat pumps energize the four-way valve in the cooling function. One of the mercury switches within the body of the thermostat has the function of energizing the four-way valve just before the heating or cooling mode of operation is needed.

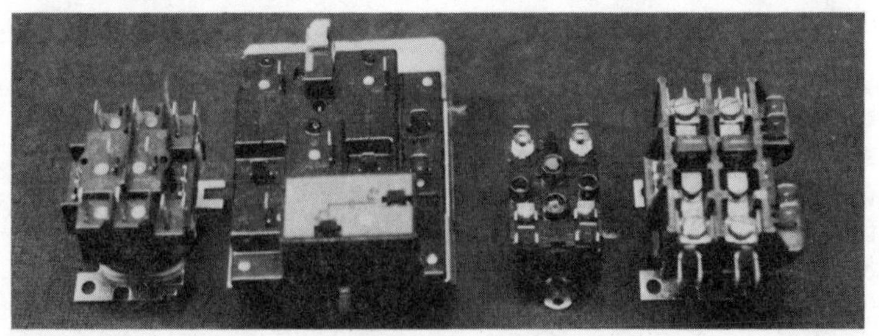

Fig. 5-4. Thermostats are basically switches, and they usually apply control voltage to relays. From left to right: double-pole, double-throw, general-purpose relay; electric heat sequencer; double-pole, double-throw, light-duty relay, and a heavy-duty relay known as a contactor.

Manufacturers claim that an additional 10 percent savings can be obtained by using new electronic heat pump thermostats. In the past we always recommended that a heat pump thermostat not be turned back at night. The new electronic thermostats compute the morning warm-up time (the time needed to heat a space) and lock out the secondary heat source in colder climates. Locking out the secondary heat source (backup heat) means that additional savings can be attained by using an automatic setback feature. Although the new thermostats are expensive, the payback ratio makes them worthwhile investments.

Heating and Cooling Automatic Switchover Thermostat

Automatic switchover thermostats are fairly new to some manufactured equipment. Automatic means that the cooling or heating function will turn on as needed, depending on the space temperature. In the past the heating or cooling function had to be manually switched at the thermostat in all cases. In the automatic thermostat the temperature adjustment levers, when set together, maintain a 3° or 4° minimum separation so that heating and cooling won't operate together. If the levers that set the temperature are next to each other, and the heating lever is set at 74° F., the following will occur. As long as the temperature is below 74° F., the heating function will operate. When the temperature is above 77° F., the cooling function will operate. Remember the 3° to 4° separation in the lever settings. When looking at the front of an automatic switchover thermostat, the cooling lever will be on the right of the heating lever (Fig. 5-5).

Heating Reactive Thermostat

When temperature causes the R and the W terminal to connect

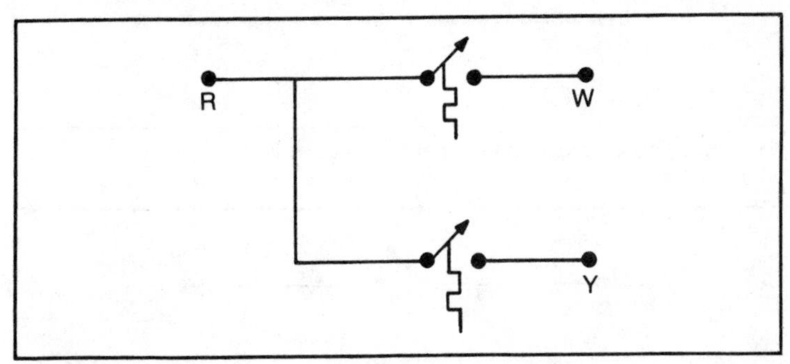

Fig. 5-5. Switching diagram for a heating and cooling automatic switchover thermostat.

through the switch, a heating function is started. Using these terminals means that when the temperature falls within a living space, R and W will make an electrical circuit. This type of thermostat is known as a heating only type or a thermostat that closes the switch when temperature falls (Fig. 5-6).

Cooling Reactive Thermostat

When temperature causes the R and the Y terminal to connect through the switch, a cooling function is started. Using these terminals mean that when the temperature rises within a living space, R and Y will make an electrical circuit. This type of thermostat is known as a cooling only type or a thermostat that closes the switch when temperature rises (Fig. 5-7).

A heating thermostat closes the circuit when the temperature of the living space falls. A cooling thermostat closes the circuit when the temperature of the living space rises. Combination thermostats switch between heating and cooling either manually or automatically.

Heating and Cooling Thermostat with a Manually Switched Fan Circuit

Most heating and cooling thermostats include a manual switch to activate the forced air fan in the indoor system (Fig. 5-8). When a heating function is needed, the temperature within the furnace will turn on the furnace fan (not a heat pump installation, just a furnace with an air conditioner). When the cooling function is started, the forced air fan must start at the same time that the air conditioning

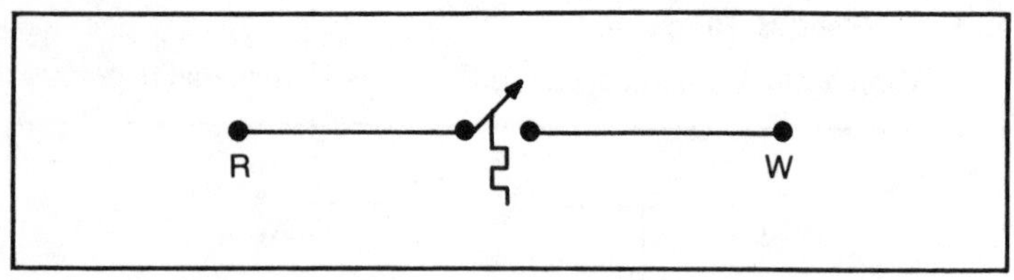

Fig. 5-6. Switching diagram for a heating reactive thermostat.

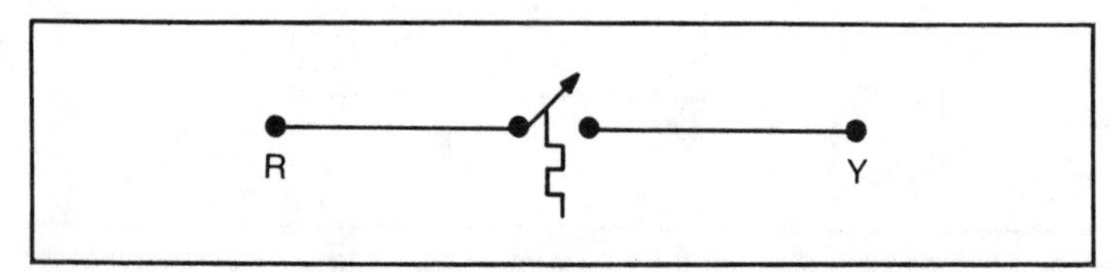

Fig. 5-7. Switching diagram for a cooling reactive thermostat.

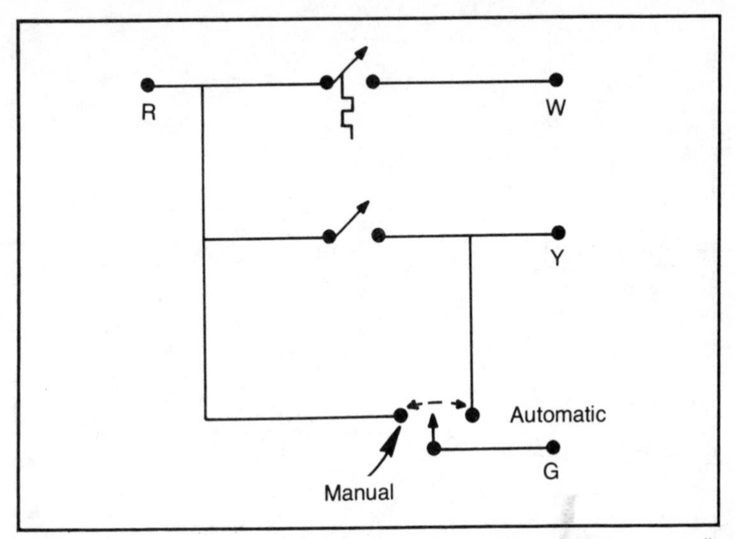

Fig. 5-8. Switching diagram for a heating and cooling thermostat with a manually switched fan circuit.

unit starts. If the operator desires to operate the forced air fan continually, then the fan switch can be set from the automatic position, to the on position. During hot weather, it's usually best to operate the forced air fan continuously to mix the indoor air.

Multistage Heating and Cooling Thermostat

Multistage thermostats have two stages of heating or cooling (Fig. 5-9). This type of thermostat is supplied in two-stage cooling, single-stage heating, two-stage cooling and two-stage heating, or single-stage cooling and two-stage heating. The additional stages are nothing more than temperature reactive switches that turn on a specific function at a manually set temperature. A multistage thermostat has a mechanical electrical offset built into it that separates the first and second stage by 1° or 2°. If the first stage, being heating or cooling, isn't able to carry the need of the living space, then when the temperature variation becomes 1° or 2° more than the first-stage setting, the second stage will turn on the proper circuit. Staged heating and cooling provides cheaper, more accurate temperature control in a living space.

ISOLATION OF A TWO-TRANSFORMER CONTROL SYSTEM

Figure 5-10 shows a method of isolating the control transformers of the oil furnace primary control and that of the heat pump.

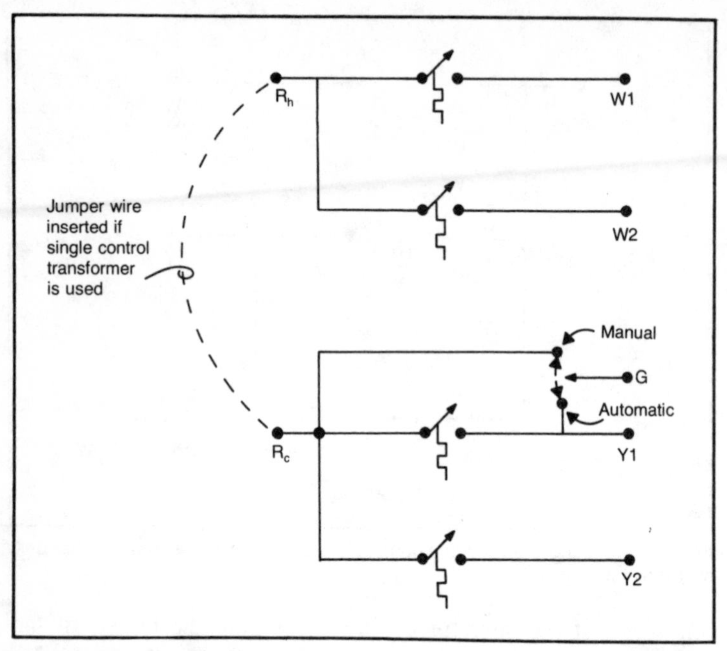

Fig. 5-9. Switching diagram for a multistage heating and cooling thermostat.

Note that the indoor coil is in the return duct to make this circuit effective. Normally an oil furnace is the only type of heating plant needing this type of control circuit. Other types of heating plants will use only one control transformer of 24 volts ac, 40-VA output.

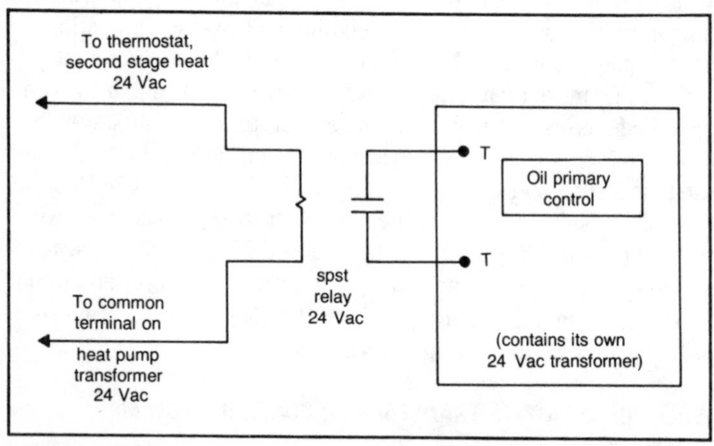

Fig. 5-10. Isolating the control transformers of the oil furnace primary control and the heat pump.

Oil is an exception because the transformer isn't replaceable within the primary control. The VA rating of the oil transformer is too small to carry the heat pump load (VA = volts times amps of the output of a transformer).

EMERGENCY HEAT SWITCH OPERATION

A switch located on the thermostat face turns the heat pump off and allows only the furnace or backup heat to warm the living space. If a problem develops within the heat pump, such as no heating or noise, the switch should be moved to the emergency position. When the thermostat is put into emergency, it deenergizes the heat pump control circuit and leaves the 230-volt power supply intact. Never turn off the power (230-volt circuit) supplying the heat pump. If possible, use the emergency heat switch to disengage the outdoor section. A small electric heater powered by the 230-volt circuit must remain functional to protect the compressor from damage when the unit is restarted. If the 230-volt supply circuit is turned off for any reason for more than a short period, a compressor reheat time period will be required. The normal reheat time for a compressor is 24 hours after the 230-volt supply is reestablished to the outdoor section or compressor section of the pump. The emergency heat switch in the emergency position (EM HT—emergency heat) will permit backup heat operation only, thus keeping the compressor section off.

A warmed compressor may load up with liquid refrigerant in certain outdoor temperature conditions. If the compressor isn't heated and it tries to pump the liquid, damage within the compressor will result.

A side effect of emergency heat operation is that the air handler or furnace fan will run continuously. The heat source will turn on and off as heat is needed in the space.

Chapter 6

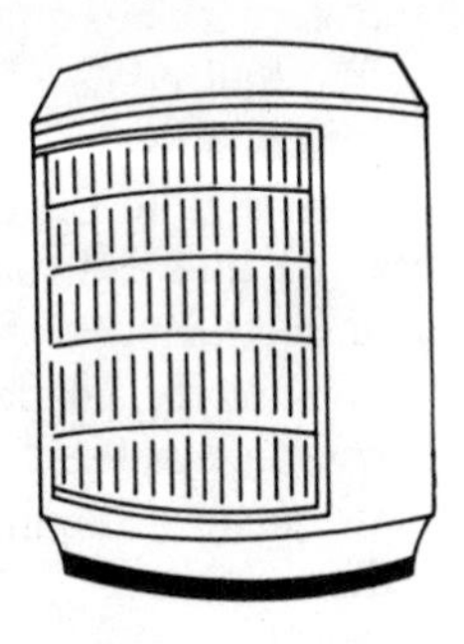

Installation Checklist

Your ability will determine how much money you will save by installing the pump yourself. Work with your heat pump dealer so you won't void your warranty. Most items in the checklist are noncritical. Do not proceed with items 27 through 34, however, unless you know what you are doing. If you are unsure, leave items 27 through 34 to a trained serviceman.

The average heat pump job is marked up $1,000 to $2,000. The markup covers labor, warranty, and profit. The more items you can take care of on the list, the better off you are regarding total costs.

PROCEDURE

1. Become familiar with all the written material.
2. Unpack all boxed sections of the heat pump.
3. Visualize where each part of the system will be installed.
4. Measure and draw a sketch of your duct system to install the indoor coil.
5. Take your drawing to the sheet metal shop for fabrication of any duct size changes required.
6. Install the newly fabricated duct and coil case.
7. Pitch the coil drain pan toward the drain slightly to allow for drainage.
8. Select a site for the outdoor section of the heat pump.
9. Install a snow curb on the site of the outdoor section if needed.

10. Install the outdoor section on the snow curb.
11. Drill a hole through the outside of the house to allow for passage of the refrigerant lines and control wire. A 2″ to 2½″ hole is usually required.
12. Drill a hole through the outside of the house to allow for passage of the power wire to the outdoor unit.
13. Uncoil the refrigerant line and rough it in between the outdoor and indoor sections.
14. Using PVC pipe, install the condensate drain line. Remember to pitch the pipe at least ¼″ per foot.
15. Install a condensate pump if required.
16. Remove the old thermostat.
17. Using the old thermostat wire, pull the new thermostat wire through the wall.
18. Mount the thermostat subbase in position where the thermostat wire penetrates the wall. Make sure you follow the directions.
19. Level the thermostat.
20. Make a note of which wire color goes to what terminal letter. Keep the note for future reference.
21. Mount the thermostat on the subbase.
22. Plan and route the thermostat wire to the air handler or to the furnace area.
23. Route the control wire to the outside unit from the air handler or furnace. It's okay to run the control wire outside by using the refrigerant pipe as a guide and support.
24. Call the electrician and make arrangements for power wiring the outside unit. Leave all the switches off.
25. Support the refrigerant lines with 2″ wide straps and foam as shown in the manual.
26. Connect the control cable from the furnace to the low-voltage terminal board in the outside unit. Connect the red wire to the R terminal and the yellow wire to the Y terminal. Make a note of which wire goes to what terminal. You will need this information later when you're connecting wires at the air handler or furnace.
27. If you know controls and understand the function of each wire in the control system, proceed to the next step.
28. Interconnect the control wires from the outside unit, the thermostat, the furnace or air handler, and the fan relay.
29. If you are using a condensate pump, wire it to the proper voltage circuit.

30. Remember to leave all switches turned off.
31. Connect the refrigeration lines to the outdoor and indoor units.
32. Double-check all your work.
33. If your heat pump has service valves. Make absolutely sure that the valves are open. Failure to open the service valves could result in bodily injury.
34. Most if not all manufacturers want the high-voltage circuit turned on for a 24-hour period before starting the heat pump. Set the thermostat on EM heat (emergency heat) and turn the line voltage on to the outside unit. If the low voltage is wired correctly, the heat pump will not start, but the compressor heater will have a chance to warm up for 24 hours. Another way is to leave the furnace turned off, set the stat for EM heat, and turn on the outside unit as far as the 230-volt circuit is concerned. Don't start the outside unit for 24 hours. By setting the thermostat on EM heat, the outside unit won't have low-voltage power and thus can not start if wired correctly.
35. If possible, call the dealer that sold the pump to you and arrange to have the unit started and checked.

CAN YOU FINISH THE INSTALLATION?

After reading the material in this book, you should thoroughly think over the installation project. Decide which parts of the job you feel comfortable in starting or completing. We feel that at the very least you can save $100 if you only do the nontechnical work. If you have a mechanical and electrical background, you can save $400 to $900. An installer working in your home will charge $20 to $40 per hour. The main areas that you will be concerned with are listed below.

- Refrigeration: you will have to know how to unroll the premanufactured tubing, how to hang the tubing in the support/vibration isolators, and how to use a wrench to tighten the couplings. You won't start the heat pump yourself, so you don't need to know pressures versus temperatures, etc.
- Electrical: control wiring and power wiring are involved. If you have done wiring jobs, then you will know wire sizes, fuses, and enough about electrical codes to do the job right and safely. Three wires have to be run from the fuse panel to a disconnect box outside, then to the unit. You will need

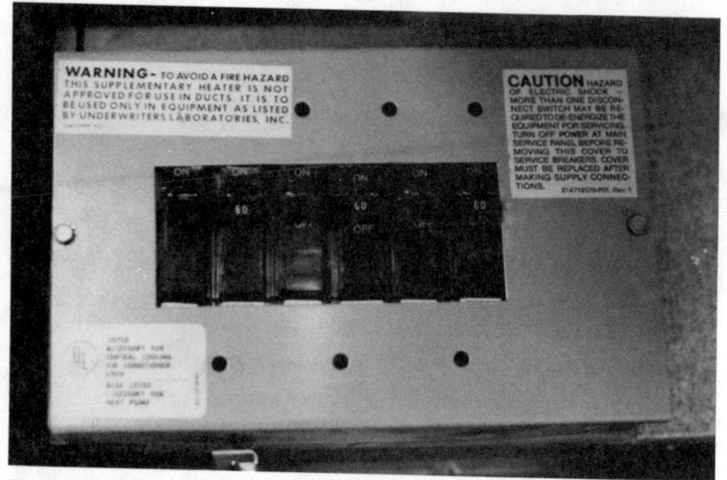

Fig. 6-1. The circuit breakers shown are part of an electric furnace. Note the safety signs on the cover plate of the circuit breaker box.

an electrical background, plus a knowledge of relays and low-voltage circuits, to do control wiring. If you are shaky with the control wiring, just run the cables and let the heat pump technician connect the wires to the proper terminals (Fig. 6-1).

- Sheet metal: the sheet metal will have to be cut to size to accept the coil housing and connectors. The drawings in this book will allow you to do this if you have a minimum of experience. Visualize where the parts will fit and cut the sheet metal to the proper size.
- Refer to the Tool List section on Chapter 4. You will need to know how to use each item on that list.
- Do only what you feel confident in doing. Leave the rest to your technician.
- Know your limitations in a technical sense.
- Visualize and know where each part will fit into place.
- Begin the work and follow your game plan.
- When you have gone as far as possible, call your serviceman to check and start the heat pump.

It is very important that your serviceman starts the heat pump and checks its operation. The serviceman will check the different functions, do a safety check, and only then will he warrant the product.

Chapter 7

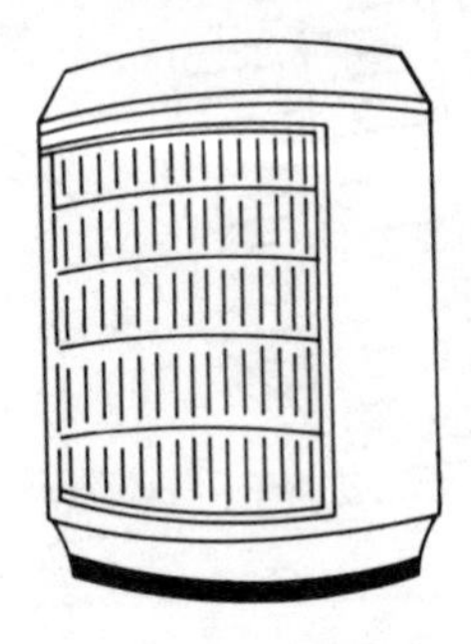

Ground Frost and Heat Pump Support Structures

Each year approximately 96,000 cubic miles of water evaporates from the surface of the earth. If we took the total rainfall in the world and dumped it on Texas, that state would be covered by 475′ of water at the end of one year. Water is in the ground, in the air, and all around us.

Groundwater is referred to as *ground frost* when frozen. If the ground remained either in a frozen state or permanently unfrozen, ground movement wouldn't present a problem regarding heat pumps. When frost leaves the ground and the ice becomes water, ground movement occurs. Whether this movement is an actual push upward or a depression in the surface is immaterial, but what is important is that the earth's change affects the heat pump minimally.

Ground frost can extend from the surface to a depth of several feet, depending on the yearly average temperature at a particular site. One way of determining the frost effects in your area is to ask your local plumber. Don't be surprised if that average depth occasionally changes by a few feet. Plumbing normally is placed below the frost line to reduce freezing and breakage.

When the outdoor unit of a heat pump system contains the compressor, level mounting can be important. The compressor uses oil for lubrication. Because the oil is kept in the base of the compressor, the base of the condensing unit should be kept near level (Fig. 7-1). The base can tip 5° to 8°. Each manufacturer has a recommendation regarding acceptable tilting, but we've found that a

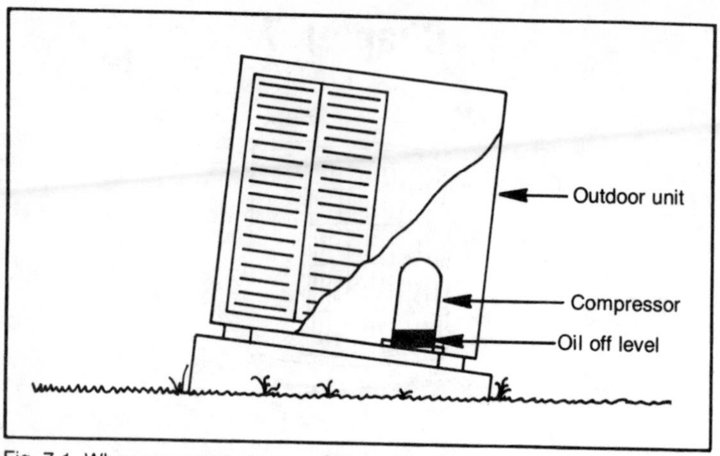

Fig. 7-1. When a support structure fails under a heat pump, severe tilting may occur. Lubrication failure might be the result within the heat pump compressor in some instances.

few degrees one way or the other are not a problem in everyday usage. If you can see that the outdoor unit isn't level, the condition should be corrected. Unless you have a very critical eye, a tilting condition of 1° or 2° isn't readily noticeable to the unaided eye.

GROUND PREPARATION

When a new home is built, settling will occur under a heat pump pad unless the ground under the pad is carefully prepared (Fig. 7-2). Even with good ground preparation, a noticeable sink rate is present in some installations. The only way to avoid ground compression under a heat pump is to allow enough time for the ground to stabilize. If enough time can't be allowed, then tamping, soaking with water, and backfilling of the surface with sand and gravel will give satisfactory results. If you have certain types of sand or rocky soil, settling won't be a concern.

If you are in doubt about ground preparation at a certain site, ask a building contractor in your area. He will know how to handle different soil compositions. Many times it's possible to locate the heat pump condensing pad on undisturbed ground. The outdoor unit shouldn't be near a bedroom, and the location has to be central to the refrigerant piping and electrical supply. Design limits us to locating the pump 50′ to 75′ from the indoor unit when an air-to-air condensing unit is placed outside.

When installing a heat pump condenser pad (sometimes referred to as a snow curb or base) at an older home, the procedure becomes somewhat easier. The best method entails digging a hole approximately 1′ larger on all sides than the planned heat pump pad or snow curb. The depth of the pad hole ideally will be 8″—enough room for 4″ of sand and 4″ gravel to act as the support base for the pad itself. Sand and gravel act as a broad support for the snow curb and also provide a drainage area. The heat pump will remove frost from its outdoor coil by a defrost cycle function; the result is water running out the outdoor unit. Special soil conditions will require some determination on your part to achieve the desired drainage in your application of the pad system.

SUPPORT PADS

Heat pump support pads can be built in any shape as long as the support requirement is met. To provide adequate support, the pad must be slightly larger than the base of the heat pump condensing unit.

Support pads can be constructed successfully from readily obtainable materials like concrete, cement blocks, formed plastic, bricks, and angle iron. When given a support base of sand and gravel concrete will give excellent results. When the site has been prepared and is level, build a form to retain the concrete as it cures (Fig. 7-3). The form should be built from 2″ × 6″ lumber to allow enough depth for the pad to remain in one piece. (Thin concrete has a tendency to crack if less than 4″ thick.) Scrap pieces of wire may be incorporated within the form before the concrete is poured to give additional strength within the concrete itself. After the concrete has cured, remove the form. You have a heavy-duty support for the heat pump. Make sure that the top of the concrete form is level in all directions. When the form is filled, you will have a level heat pump base.

A disadvantage of a concrete support pad is its heavy weight. If a frost heaving condition occurs and pad tilts, lifting the pad back to a

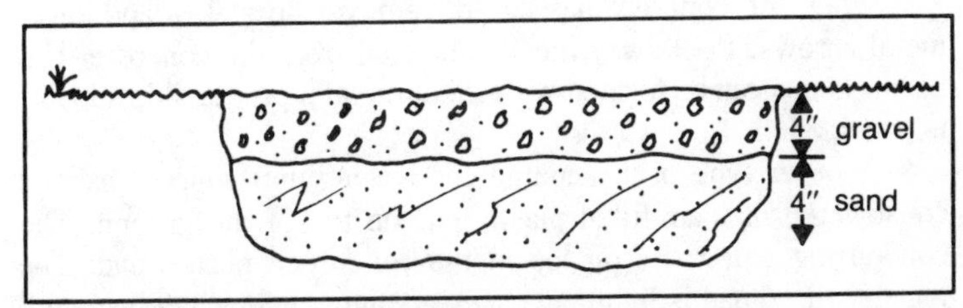

Fig. 7-2. Ground preparation is an important step in building a base for the heat pump support pad. Geographical areas with a very low water content in the soil won't need a sand and gravel backfill.

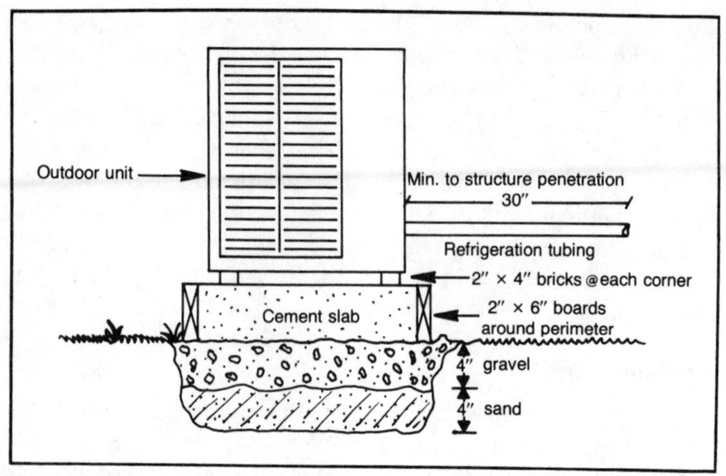

Fig. 7-3. Concrete supports are easily poured within a 2 × 6 wooden form.

level position is very difficult. Nevertheless, a concrete support is the most solid support if it is installed properly.

If you require a concrete base that is round, it's possible to cut a section of sheet metal for the height and circumference of the circular shape desired. Roll the cut sheet metal so that the ends meet. Attach one end to the other with sheet metal screws, thus forming a circle. Mark a circle on the ground where the concrete pad is to be installed that is the same size as the sheet metal concrete support. Place the sheet metal on the circular marking on the ground. (The circle marked on the ground acts as a reference point for the sheet metal.) When the sheet metal form approximately matches the circle on the ground, hammer wooden or metal stakes into the ground to provide vertical support for the sheet metal form (Fig. 7-4). Level the upper surface of the form. The top of the form will act as a reference level for the top of the concrete when it's being poured.

When the concrete has cured, remove the stakes and sheet metal screws. Peel away the sheet metal from the concrete. Unusual shapes can be formed with sheet metal if proper side support is provided.

Another type of heat pump condensing unit support base is constructed of foam-filled plastic put on top of the ground. The condensing unit rests on top of the reinforced plastic pad. The manufactured pad is light and easily applied. Unfortunately, a condensing unit will slide around on the top of the plastic pad unless rubber is applied between the plastic material and the heat pump

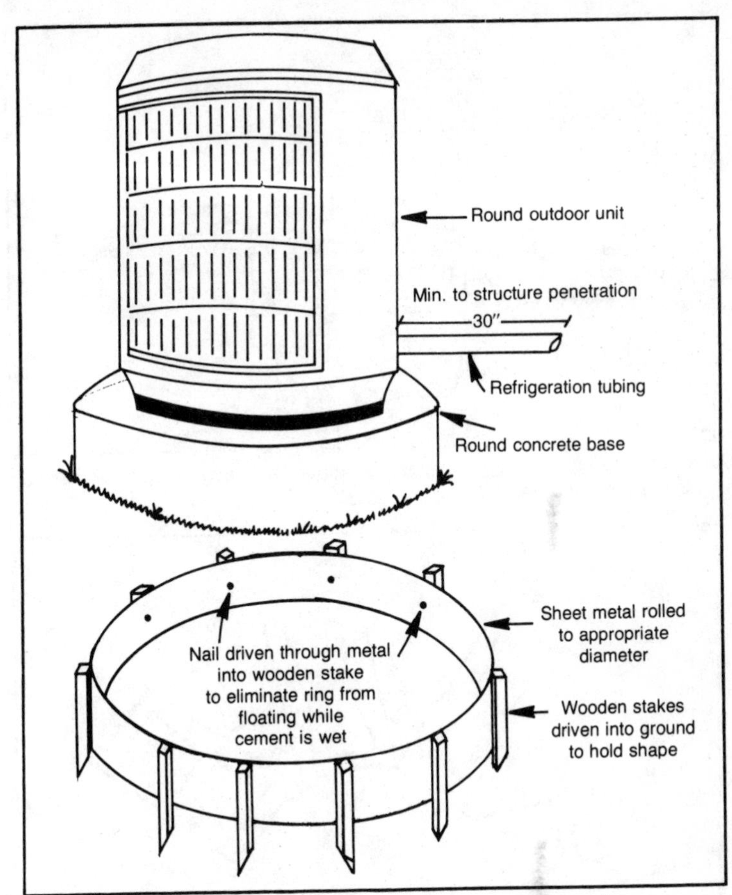

Fig. 7-4. Round heat pumps can be mounted on a round snow base support.

Fig. 7-5. A simple concrete block support for a rectangularly shaped heat pump.

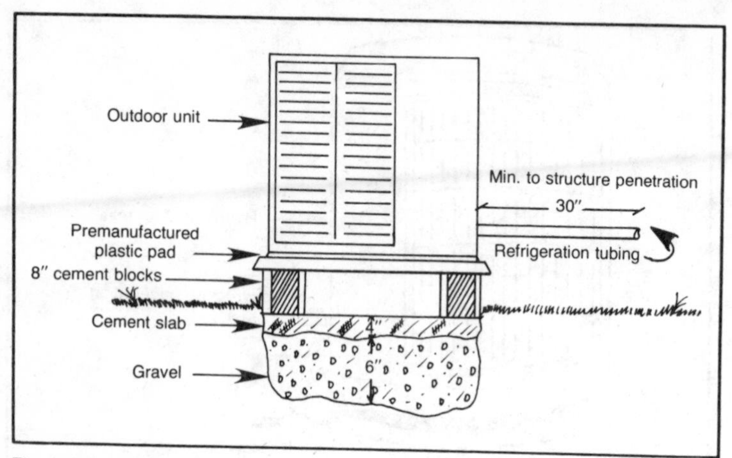

Fig. 7-6. The addition of a sheet metal cover over the cement blocks will make an installation more attractive.

Fig. 7-7. Imagination plus a few extra cement blocks will give a better overall appearance. This pattern is for making a round heat pump base for eight cement blocks.

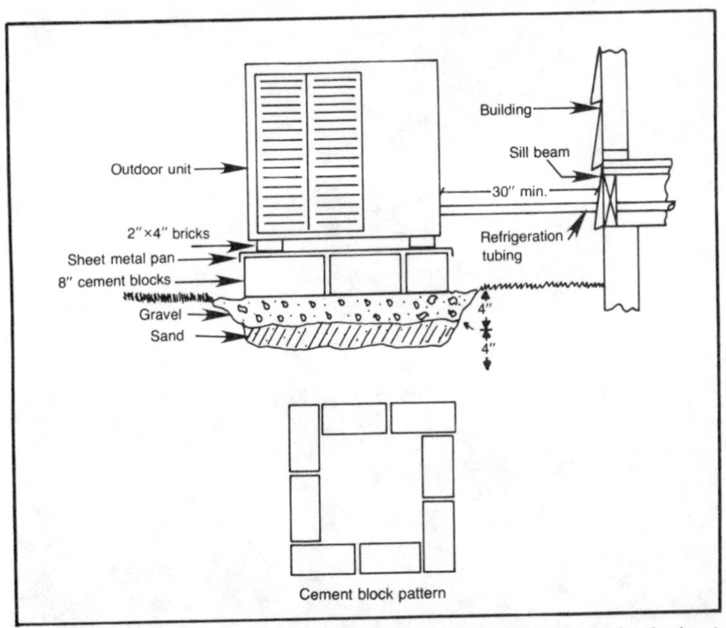

Fig. 7-8. A sheet metal cap can be added to cover the support base for the heat pump. Several different configurations can be achieved with the use of cement blocks.

Fig. 7-9. Maintain air flow by removing weeds around the outdoor unit.

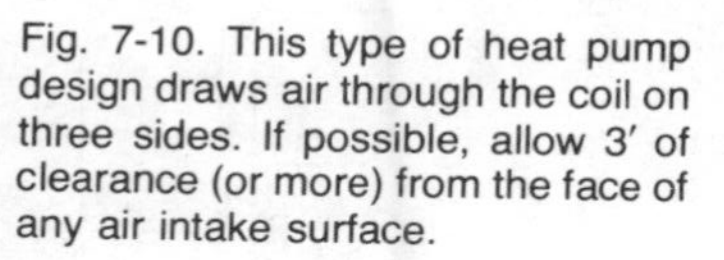

Fig. 7-10. This type of heat pump design draws air through the coil on three sides. If possible, allow 3′ of clearance (or more) from the face of any air intake surface.

condensing unit. This plastic pad is expensive. Most of the support pads in this book are made of inexpensive materials, but they do require labor.

Still another type of support pad can be constructed of concrete building blocks. These blocks are inexpensive, strong, and ugly. The most inexpensive type of block pump support consists of just four blocks set at each corner of the outside unit (Fig. 7-5). (The

Fig. 7-11. Higher than average snow bases can be constructed from angle iron. Note the steam rising from the heat pump. The outdoor air is cold, and the unit is defrosting.

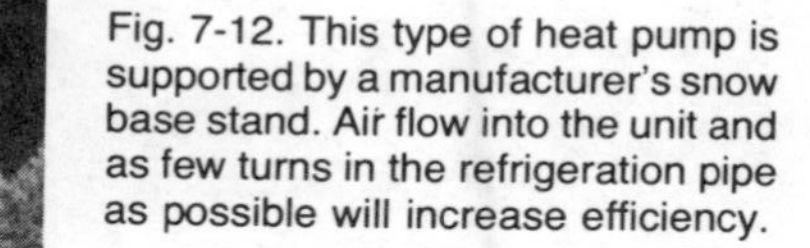

Fig. 7-12. This type of heat pump is supported by a manufacturer's snow base stand. Air flow into the unit and as few turns in the refrigeration pipe as possible will increase efficiency.

Fig. 7-13. Notice the sweeping turn of the refrigeration piping. The thermostat wire is secured to the piping (line set) by ordinary duct tape. The line assembly is attached to the wall with sheet metal straps insulated with foam rubber.

Fig. 7-14. An electrical disconnect box is mounted on the wall behind the outdoor unit of the heat pump.

outside unit and condensing unit are two names for the same piece of equipment.)

With the application of a little American ingenuity and a few more cement blocks, an unusual attractive support pad can be constructed (Figs. 7-6 through 7-14). We stress the use of sand and gravel so that you'll have a solid base for your heat pump.

Chapter 8

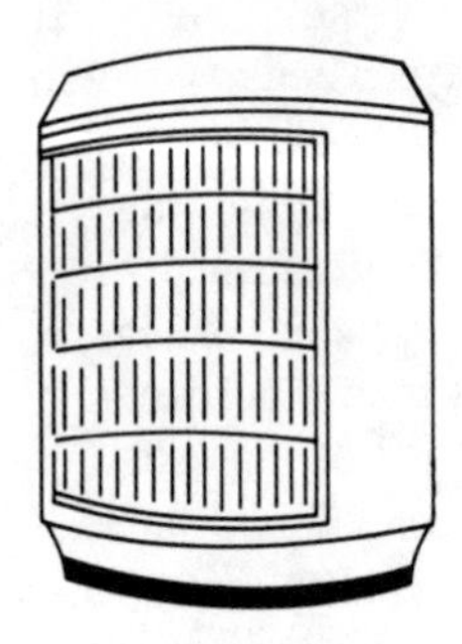

Service Information

Follow these safety tips and use your intelligence. Don't experiment with valves unless you know exactly what the results will be within the machine or to yourself.

- Never use a torch or flame on a charged refrigeration line or part.
- Never cut into a refrigeration line or part unless you know that it doesn't contain pressure.
- Never close any valve unless you know what the effect will be on the machine and yourself.
- If you feel uncertain about any of the procedures, refer to a service technician or the equipment instructions. Work within your ability.
- Never put a stick or any object into a moving mechanical device.
- Never work on a piece of equipment unless all power is off. Turn off the entire system. This includes the furnace if you're working on the outdoor unit.
- Wear safety goggles.
- Remember that the oil within the tubing may ignite when a torch is used. Have a fire extinguisher handy.
- Imagine what could happen to hurt you as you work and exercise caution.
- Be careful of sharp metal edges.

SERVICING YOUR HEAT PUMP SYSTEM

A heat pump has a "personality" in its own right when compared to other types of heating equipment. An ordinary furnace expends maximum energy when it's operating. The furnace doesn't release more energy if the thermostat setting is lowered on a single-stage furnace. Multiple-stage, electronic or electromechanically-controlled units furnish more then one level of heat output, depending on the amount of heat required in a space. These furnaces comprise less than 1 percent of the marketplace and are confined to gas or electric heating units in residential applications.

The average air conditioner in a residential application is either totally on, as far as output is concerned, or completely off. Multistage units are on the market, but because of cost they are in few applications. If you only need 25,000 units of heating or cooling, it would be better to supply only that amount to the conditioned space.

Heat pumps have a quality in their heat output that is quite different from that of a furnace. The colder the outdoor temperature, the less heat the heat pump has to deliver to the inside of the residence. This is true with air-to-air pumps or even water-to-air heat pumps. The water-to-air pumps have a higher overall output and are more expensive to operate.

Because of the reduced output in air-to-air heat pumps, the operating time extended to produce the heat required by the home. Better service is needed to maintain the pump in the best operating condition to attain the best efficiency.

Air Filters

Air filters are the simplest and most costly of the service items to overlook. The heat pump has to maintain heat transfer to be efficient. In the heating mode, as the filter plugs up with dust and lint, the ability of the heat pump to transfer heat is reduced, then a pump failure occurs. This efficiency is costly in the long term, but a pump failure can be very expensive because of the increased mechanical wear that will occur.

Electronic air filters can remove 95 percent of the airborne particulates, including smoke and cooking grease that is suspended. Paper or composition filters provide better filtration through increased density and greater surface area. The cost of an electronic filter is about $350 installed, and it only needs to be washed to maintain its efficiency. The high-efficiency composition filters cost from $15 to more than $100. They do require replacement. High-efficiency filters are considerably less efficient than the electronic types. Fiberglass furnace filters are inexpensive, easy to change, and are available almost everywhere. The efficiency of fiberglass filters is comparatively low—3 to 10 percent. Fiberglass filters work fine if they are serviced as needed.

Sources of air filter problems include carpets, leakage from clothes dryers, household pets, and outside air leaking into the home. The filter should be changed once a month without fail. If you live in a reasonably tight home and have no children or pets, you only need to change filters once every three or four months.

If you are using a new fiberglass or composition filter, hold it up to a light and note the amount of light coming through it. You might be able to see through the filter when holding it up to a light bulb. After 30 days, remove the filter and hold it up to the light. If the amount of light coming through it is noticeably less, change the filter. Scanning the surface of the filter may help you determine the sources of your dust or lint.

Vacuuming is acceptable for fiberglass or composition filters, but not for electronic air filters. When vacuuming the filter, take care not to pull holes into the filtration material. When the filter is vacuumed, it should be coated with one of the dust mop coating aerosols on the market. The aerosol increases the filter's ability to pick up and hold dust that strikes its face. Heating equipment dealers usually carry special filter sprays meant for this purpose, but we're not convinced they are better than the grocery store mop spray. Electronic air filters are never vacuumed because they are susceptible to loose particles falling between the plates. Electronic filters collect much smaller particles than the other filters. Washable, nonelectronic air filters have nearly the same efficiency as fiberglass units, but you don't have to continually buy new filters.

After you have checked the filters a few times, you'll know how often to change or clean the filtering system. You must keep reasonably clean filters in your system. The dirtier the filter, the less heating and cooling are transferred to the duct system, and the higher the refrigeration pressure is inside the piping and compressor. Too much refrigerant pressure induces abnormal mechanical wear in the compressor. The small line going to the outside unit should never become hot to the touch. The temperature on the small line is a direct indication of the efficiency of temperature removal. Compare the temperature of the large line to the small line for a graphic illustration of the effect. After about 10 minutes of operation, the large line should be so hot that you can barely touch it. It brings the accumulated heat into the furnace. The refrigerant flow then returns to the heat pump through the small refrigeration line in the heating mode. If the temperature is too high on the small line, you have an air flow problem. Air filters, closed registers and blocked return grilles are common causes of this lack of efficiency in your system.

Indoor and Outdoor Coils

The indoor and outdoor coils of your system may require cleaning if they are clogged. The amount of air that the heat pump uses has been predetermined by the manufacturer. If that amount of air is reduced for one reason or another, efficiency of the system drops.

If the outdoor coil becomes restricted during heating operations, the available heat in the outdoor air isn't extracted because of reduced air flow through the outside unit. It doesn't really make any difference whether the outside air flow is restricted by a bush, dirt in the coil, or even snow. The restrictions must be kept to a minimum, because the amount of air directly relates to the amount of money you will save in operation costs.

In the cooling mode of operation, a dirty outdoor coil can cause equipment damage because of increased operating pressures within the system (Fig. 8-1). Instead of picking up the outdoor heat, the system is getting rid of the indoor heat through the outdoor coil. If the indoor heat is trapped within the system because of a dirty outdoor coil, the pressures will increase in the refrigeration system. Remember the basis for refrigeration and air conditioning—temperature and pressure are directly related. If the pressure of the refrigerant increases, it's been exposed to a higher temperature. Conversely, if the pressure of the refrigerant becomes lower, it's because of a lower operating temperature, assuming that a mechanical malfunction hasn't occurred. Air flow, temperature, and the refrigerant pressures are linked in a way that affects system performance.

Air Flow and the Outside Unit

Air flow into and out of the outside unit is extremely important. The placement of the outside section of the pump shouldn't restrict the air flow in any way, or performance of the unit will suffer. Ideally, 4′ of clearance should be maintained around the pump in any horizontal direction. This horizontal distance assures that the outdoor unit air will be available in the quantity needed. If your unit

Fig. 8-1. Follow safety rules when washing a coil.

needs service, adequate space has been provided. If there is any question, refer to the manufacturer's instruction packet. If weeds, flowers, or shrubs block the flow of air, the system will be less efficient. It takes power to move air. When the air movement is restricted, it takes additional power to move the same quantity of air. The additional power isn't available in residential heat pumps, so the air flow suffers. If you can maintain a distance of 4′ around your outdoor unit, you shouldn't have any problems with the air intake to the system.

Most outdoor unit designs today are using horizontal intakes with the discharge air leaving the unit in a vertical direction. Some noise is a by-product of operation of the outdoor unit. We have noticed that in the vertical discharge systems, the majority of the fan noise seems to be carried away with the air leaving the outdoor fan.

Never place the outside unit in an enclosure or allow the air leaving the unit to strike any object within 8′. If a shrub grows over the outdoor unit and disturbs the discharge air flow, the exiting air could start a rolling motion. The rolling motion would cause the processed air to slow too near the unit and possibly recirculate, lowering the overall efficiency. We've seen instances where a homeowner built a roof over the heat pump or air conditioner

Fig. 8-2. This debris should be removed from the outdoor unit.

outdoor unit, thinking that it would keep the rain or perhaps leaves out of the unit. The outdoor units are made to be outside in the weather. If leaves or needles accumulate in the outdoor section, they should be removed manually.

Inspect the outdoor coil to ensure that the air flow passages are clear of contamination. The fins on the outdoor and indoor coils are spaced at less than 0.1″ apart. Contamination will usually layer in a thin deposit on the air-entering side of the coil (Fig. 8-2).

Cleaning of the coil surface can usually be done by just brushing in the direction of the fin slots (Fig. 8-3). This brushing has to be done carefully. If the fins become bent, the effect is to reduce the air flow through the unit. Brush gently in a straight even movement in the direction of the slots. Turn the power off to the unit or units when doing any service procedure.

If brushing the outdoor unit doesn't clear away the clogged condition, other methods are available. With the power turned off to the indoor and outdoor units, it's possible to use a pressure washer to spray the coil with detergent. Do not allow water to enter the outside fan motor housing. If you spray downward at an angle of 30° to 45° this method works quite well. If water enters the outside fan motor, the motor may burn up. We do not recommend this method if you're unfamiliar with electrical safety.

If you wash the coil, rinse it thoroughly because the fin material may react to the cleaner and decay. If your cleaner can be used on

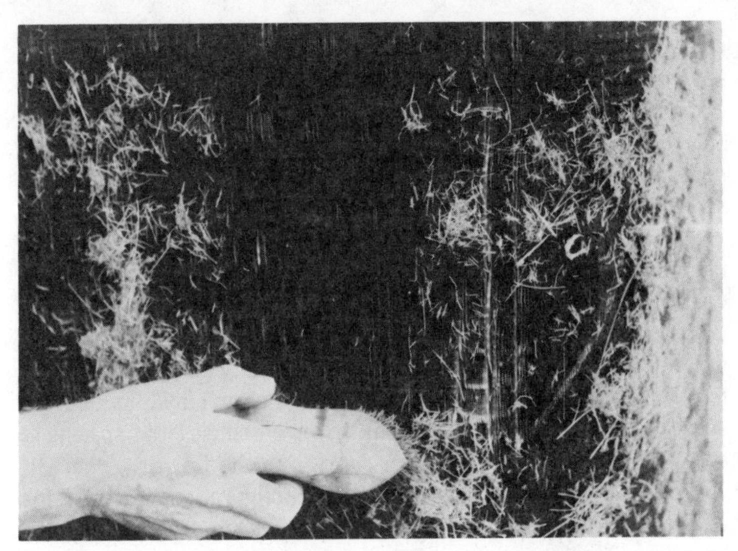

Fig. 8-3. Brushing the fins of an outdoor unit.

aluminum, it should be okay. Some soaps are easier to rinse than detergent. They are biodegradable and somewhat inert chemically.

If the contamination in the outdoor coil isn't greasy, use a hose and water (Fig. 8-4). Be sure and maintain the 30° to 45° angle to

Fig. 8-4. If the spaces between the fins are clogged with material, wash the coil.

protect the fan motor. If the fan motor becomes wet, let it dry for several days. Covering the fan motor with plastic or removing it is a good idea.

Another method of cleaning the coil is to use compressed air on nongrease-based dust accumulations. An air compressor would be safest for the equipment.

If a few coil fins become bent during cleaning, a fin comb can be used to straighten them. It can be purchased at your heating and air conditioning dealer. The density of fins per running inch varies, depending on the manufacturer. Your coil may have 10 fins per inch or 14 per inch. The number of fins doesn't really make a difference, but the slots of the fin comb must match your coil fins. If you use a comb that doesn't match the coil fin arrangement as far as spacing is concerned, you will cause more damage to the fins than you will correct. Fin combs are usually sold in configurations allowing for different fin densities in one tool. Notice in Fig. 8-5 the distance between the slots in the comb. Each group of slots represents different fin densities found in air conditioning and heat pump coils.

The fin comb is inserted into the fins above the bent section. The comb is drawn down the fins and through the bent section with a smooth even force. The result is that fins are once again separated so that air may pass between them (Figs. 8-6A through 8-6C).

If the finned section is bent with considerable force, it may not be possible to return the fins to a like-new condition. As any metal is bent, it stretches. Because of the stretched condition of the metal, it's not possible to straighten it unless it's compressed to the starting size of the stamping. If you should encounter a badly bent section of fin material, just straighten it as well as you can and let it go. Hopefully the badly bent section will be only a small part of the total surface area on the coil face. Bent fins mean less air flow, less air flow means less efficiency, and less efficiency means higher operational costs.

Fig. 8-5. A fin comb is used for straightening the fins of air conditioning coils.

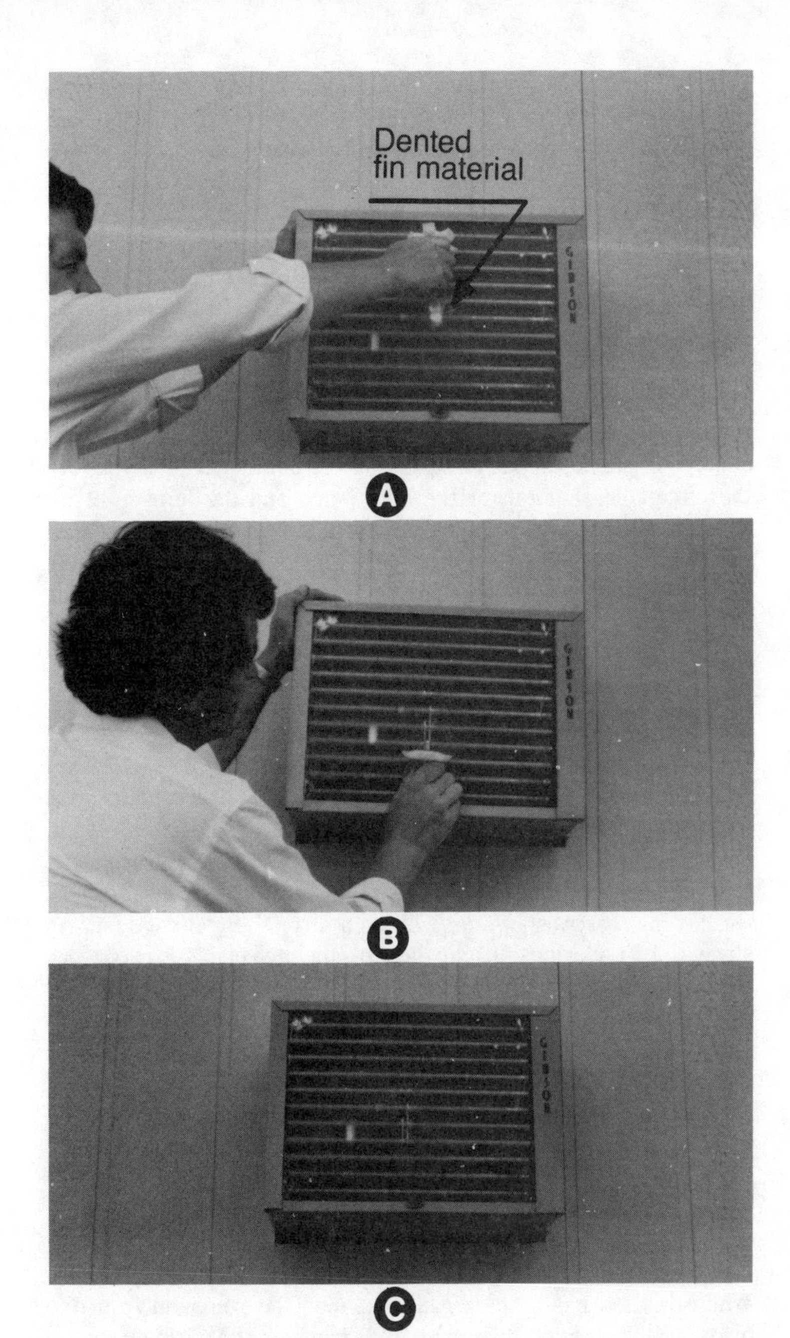

Fig. 8-6. (A) The fin comb is inserted into the fins above the bent section. (B) The comb is drawn down the fins and through the bent section. (C) The fins are separated so that air can pass through them.

All heat pumps contain a drain pan on the bottom. Water from the outdoor section will drain to the bottom pan of the casing of the heat pump. If the water was allowed to stand in the bottom of the heat pump cabinet or casing, it would cause the cabinet to rust or the ice to expand. The bottom pan of the heat pump will contain drain holes. The drain holes should be checked periodically and cleaned if needed so that the outdoor unit will drain properly. If ice builds up inside the heat pump cabinet, to more than a 1" or 2", the cabinet may warp during operation of the pump in cold weather. To gain access to the drain pan, panels will have to be removed from the pump. Make sure all electric power is turned off to the outside unit. More than one switch or circuit breaker may be required to stop all sources of electrical power supplying the outdoor unit. If you turn the circuit breaker off for the heat pump and the furnace or air handler, all the power should be cut if it was wired properly. If you're in doubt, use a voltmeter for verification. When cleaning the drain pan, don't splash water on any electrical devices within the pump.

Direct-Drive Blower

The fan blades on the indoor and outdoor sections of the pump should be checked annually for abnormal amounts of dust or contamination. The blades can be cleaned with a brush or a vacuum cleaner if the contamination doesn't have a greasy base. If oil and dirt are mixed, it may be necessary to remove the fan from the housing and wash it with detergent and water. A high-pressure washer in a car wash works well on a dirty squirrel-cage fan. As always, don't overlook turning the power switch off before you work on the equipment. Furnace blowers come in two basic types that require slightly different procedures for spray cleaning. The first type is a direct-drive blower, and the second is a belt-drive blower (Fig. 8-7).

A direct-drive blower squirrel-cage unit is a fan normally found in newer gas and electric furnaces. The fan is totally supported by the fan motor bearings, and the motor is supported by the blower housing (Fig. 8-8). The complete blower assembly will be secured to the furnace through a slide rail system or sheet metal tabs built into the fan housing, then locked down with sheet metal screws. When this fan has to be cleaned, it must be removed from the furnace. With the power turned off, disconnect the wiring from the fan (Fig. 8-9).

When the fan assembly is removed from the furnace or air

Fig. 8-7. Direct-drive blowers are compact in design, and they offer high reliability and quiet operation.

handler, the fan motor will have to be removed from the fan if you plan to use a pressure washer for cleaning. Water must not get into the electric motor winding or parts. If you use a vacuum cleaner to clean the fan, the motor won't have to be removed (Fig. 8-10).

If you remove the fan from the blower assembly, tighten the setscrew—only on the flattened section of the motor shaft. Note where the setscrew is positioned before removing the motor from the fan. If you tighten a setscrew on a rounded portion of the motor shaft, the shaft material will spread out due to the pressure of the setscrew pushing on the shaft. When this spreading occurs, it will be almost impossible to remove the fan from the motor shaft. When the setscrew is tightened on to the "flat" ground on the motor shaft, the spreading doesn't affect the close fit between the shaft and fan hub.

After the fan assembly has been cleaned, reinstall it into the furnace or air handler. Spin the fan wheel by hand and check the

Fig. 8-8. A direct-drive fan assembly.

clearances after assembly. The wheel or motor should not be misaligned as damage will result when power is applied. When the clearances are sufficient for operation, reconnect the fan motor wiring to the same points from which it was removed during disas-

Fig. 8-9. Note the number of wires below the fan assembly that must be removed from the fan before cleaning. The wiring is usually in various colors to assist in identifying the circuits.

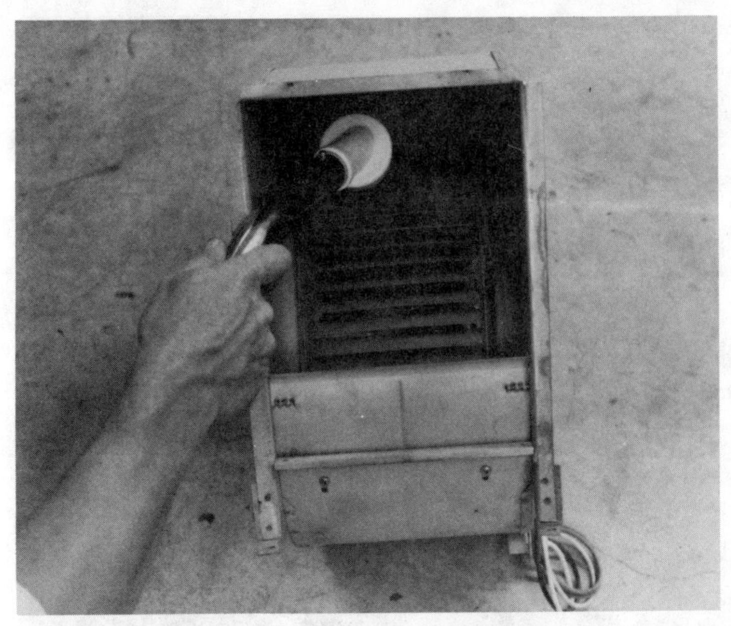

Fig. 8-10. Cleaning a fan with a vacuum cleaner is preferred. If the fan needs more than just vacuuming, the motor will have to be removed before the blower assembly is washed.

sembly. While the power is still turned off, think through the entire procedure. Make sure that you have tightened every screw or bolt on the assembly. Leaving the power switch off, go upstairs and turn on the thermostat fan switch, then go back to the furnace or air handler. Turn on the power switch for the furnace. Listen for any unusual sounds other than the smooth, even flow of fan operation. If any unusual sounds are heard, cut the power immediately and investigate.

Direct-drive blower motors are either oilable or permanently sealed. The bearings are located in the round end sections (end bell) of the motor. If the motor can be oiled, a small 1/16th-inch hole will be found directly over the bearing when the shaft is horizontal (Fig. 8-11). Often the oil hole will have a plastic or metal plug inserted into it. After removing the plugs with a pair of pliers (one plug for each bearing), add five drops of oil to each bearing. Five drops per year of 20- or 30-weight, nondetergent motor oil is recommended. Never use sewing machine oil or any oil lighter than 20-weight. Oilable and nonoilable motors look the same. The plugs or holes will be your only indication of what kind of motor you own. The motor is of the sealed type if it doesn't have any plugs. A sealed motor will

Fig. 8-11. Oiling a direct-drive blower may require removing the fan from the furnace or air handler.

last 10 to 15 years. The oilable type will last about five years longer if maintained properly.

Belt-Drive Blower

Belt-drive blowers perform the same function as their direct-drive counterparts, but the cleaning and disassembly are somewhat different (Fig. 8-12). Belt-drive units are usually limited to the lowboy furnace, but not always. The fan assembly can either be bolted, screwed, or just sitting on the fan compartment floor. As with all service procedures, disconnect the power from the unit.

A belt-drive blower consists of the blower motor, pulley system, fan shaft, bearings, and the fan. Cleaning this type of blower is somewhat easier because of easy access. If a vacuum cleaner is used, the fan assembly may not have to be removed from the furnace.

Fig. 8-12. A belt-drive blower with a heat pump coil mounted above the fan assembly. Notice the condensate drainpipe on the right-hand side of the photo.

To remove a belt-drive blower from a furnace, disconnect the power. Remove the power wires from the fan motor. Remember which wire went to what motor terminal. Check the base of the fan assembly for hold-down screws and remove them if any are present. Pull the fan assembly from the furnace so that you have easy access to all fan parts (Fig. 8-13).

Seldom is a belt-drive fan not oilable. The bearings for the motor are located on each end of the motor when the shaft is horizontal. Usually metal caps will cover the oil parts to the bearing packing. Sometimes a plug is used to cover the oil holes, but not often. After two years of average use, the motor will require oiling. Use a 20- or 30-weight, nondetergent motor oil. Five drops of oil per bearing is all that's required. Too much oil will damage the motor. The oil is held in a packing that resembles yarn. When the packing is full of oil, its only other course of action is to dump excess oil inside the motor, thus causing the damage. Oil is an insulator, but it collects dust and will turn to carbon when exposed to arcing.

Fig. 8-13. Belt-drive blowers are usually found in older residential heating units. The speed of the fan can be altered by changing one of the pulley sizes. Note the loose fan belt.

Fig. 8-14. You can measure the amount of tension on the fan belt by pulling on it. This belt is too loose.

To remove the fan belt, grasp the belt and rotate the edge of the belt over the outer edge of the larger of the two pulleys. The fan belt should be checked annually for abnormal wear or cracking. Fractional horsepower belts should last two to four years of average service on your furnace.

Excessive fan belt tightness causes the fan motor to draw too much starting current. It's also hard on the fan and motor bearings. The rules for fan belt tension follow.

- If your furnace fan motor is ½ horsepower or larger, the belt will have to run fairly tight. If the size of the motor pulley compared to the fan pulley is two to one or less, the tension will have to be tight or slippage will result. If the diameter of the motor pulley is 3.5″ and the fan pulley measures 5.75″, the ratio is less than two to one. In this case the belt will have to run fairly tight. Adjust the belt tension for ½″ to 1″ of movement from its center while on the pulleys. Note that the movement is from the center or at rest position to the point that it starts resisting further movement. Grasp the belt between two fingers to test for this movement. If the belt slips while running, it's too loose (Figs. 8-14 and 8-15).
- If your furnace uses a 1/6, 1/4, or 1/3 horsepower motor, the belt will run effectively with less tension. To adjust the

Fig. 8-15. After adjustment, the free play of the belt is reduced to an acceptable limit. The tension is right when the belt doesn't slip after the motor is up to speed.

belt properly, 2″ to 4″ of travel from center in one direction will suffice. Start the fan motor. If you notice that the belt slips for a second or two on the pulleys, you have found the correct tension. Fractional horsepower motors generally don't have the starting torque, bearing surface, or the mass to snap the fan up to speed. We are talking about split-phase motors, not capacitor-start units that were in service on the old strokers (coal burners). If you adjust the belt too tightly on a smaller motor, the results are excessive starting current, starting switch wear, and current surges in your home that may cause the lights to flicker when the fan starts. These small motors will draw about 10 times their normal running current when they start from a dead stop. A tight belt will cause starting currents that are much higher than the times 10 value. A tight belt will cause the motor to use excessive amounts of electrical power, and bearing failure will be the result. Fuses that blow out for no apparent reason are an indication of a belt tension problem.

You may question if the times 10 value for a motor-starting current is valid. Your furnace fuse doesn't blow out, and it's only rated at 20 amps. The instantaneous current that it takes to cause a circuit breaker to open is rated at 10,000 amps on many circuit breakers. Instantaneous current is not the same thing as constant current flow. Most fuses will take an overload for a period before they blow out. More information on surge currents and voltages can be found in electronic or electrical textbooks.

To remove the fan motor from the fan assembly, loosen two machine screws and remove the hold-down clamps (Fig. 8-16). It saves time if you don't remove the motor baseplate attached to the fan assembly. If the baseplate is removed, alignment procedures will be required that are a waste of time if all you want to do is spray clean the fan. The base mounting plate for the fan motor is taken off only when the motor is replaced or when fan belt-pulley alignment is needed.

Take the fan to your neighborhood car wash (Fig. 8-17). The motor and belt have been removed. The water won't damage the fan, shaft, or the bearings for the fan if you use one precaution. Don't force water into the bearings on the fan shaft. Clean the rest of the fan, but remember that the bearings are somewhat sensitive to moisture. Let the fan drip dry. Dry the spots of water that remain on the fan shaft and housing with a rag. Before you reassemble the fan assembly, oil the fan shaft bearings, plus a drop or two on each side of the bearing on the shaft. Before the motor is remounted on the

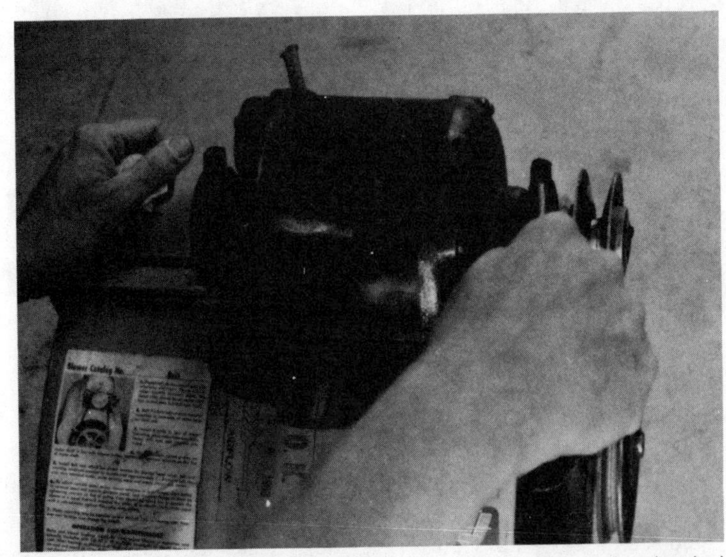

Fig. 8-16. To remove a belt-drive motor from the fan assembly, turn the electrical power off to the circuit supplying the blower. Remove the power wiring from the motor. Remove the fan belt by rolling it over the larger of the two pulleys. Loosen and remove the hold-down clamps.

Fig. 8-17. This belt-drive fan is ready for washing.

housing, the fan shaft end play will need to be checked. End play is the amount of free travel of the fan shaft, from side to side, against the thrust collars (a thrust collar is a mechanical stop collar that secures the fan shaft) of the fan shaft. Very little end play, maybe .001″ down to zero clearance, results in quiet operation of the fan. Figure 8-18 shows a fan assembly, squirrel-cage fan, fan shaft, and a thrust collar.

Remount the motor on the fan assembly. Make sure that the hold-down clamps are snug. If a ground wire was attached to the motor from the main housing, it's important to reconnect it solidly. This grounded connection is protection from electric shock if the motor should develop an electrical fault. Put the fan belt on the small pulley. With a rolling motion or rotation, reinsert it on the larger of the two pulleys. Check the alignment of the two pulleys in relation to each other (a straightedge may help you make this determination). The next step is to check the belt tension between the two pulleys. The adjustment of this tension is controlled by a height-adjusting bolt on the motor base. If you're not able to adjust the belt to the correct tension, a smaller or larger belt may be required.

Use only fractional horsepower fan belts for fan motors that are rated less than ½ horsepower. It doesn't take as much power to make the belt bend around the pulley as it functions. A fractional horsepower belt will also run quieter than, say, an automotive belt. Fractional horsepower belts are usually ½″ wide, but ⅜″ belts are

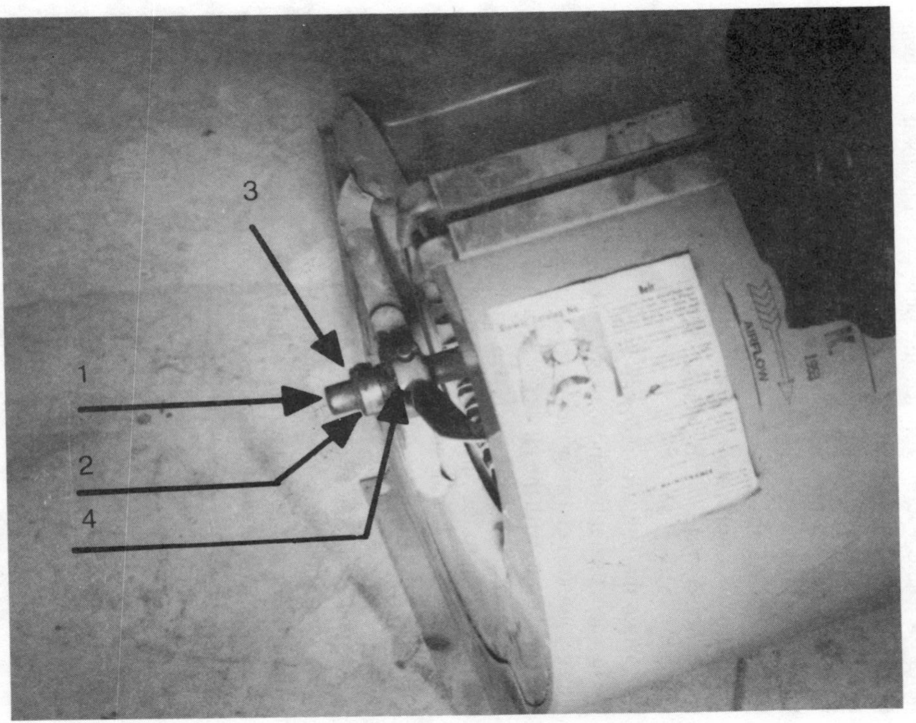

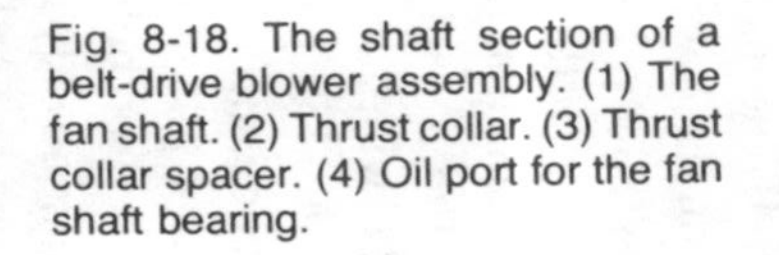

Fig. 8-18. The shaft section of a belt-drive blower assembly. (1) The fan shaft. (2) Thrust collar. (3) Thrust collar spacer. (4) Oil port for the fan shaft bearing.

in use on about 10 percent of the residential furnaces today. If your furnace has a 4L430 fan belt, 4L means that the belt is ½″ wide on the upper surface and 5/16″ on the lower surface. The number 430 means that the belt is 43″ long; the zero is ignored on all ½l series belts. The beveled edge of the fan belt is set by the upper and lower surface widths. The 4L also means that the belt is rated in fractional horsepower. Let's consider the 3L series of belts. If you have a 3L430, the upper surface is ⅜″ wide, and the lower surface is 7/32″ wide. As in the 4L series, the 430 number means that the belt measures 43″ long on the outside surface. Note that 3L and 4L belts are not interchangeable, except for an emergency.

One other popular method of belt numbering is used by a nationwide company, but you need only remember to use fractional belts the same width to match the pulleys. Your local hardware store is usually a good source for belts. If the hardware store people can't get a belt with the same number, they can at least cross match to another numbering system without the need for measuring belts. Often when belt measuring starts, if you have the number that was on the old belt, the person you're dealing with doesn't know the product. Motor horsepower ratings in excess of ½ horsepower will use either A-series or B-series belts, depending on the pulleys used.

When the belt-drive blower assembly has been reinstalled in the furnace, turn it by hand to double-check. You shouldn't hear any unusual noises such as grinding or scraping within the housing. If this checks out, with the furnace still turned off, go upstairs and turn on the thermostat fan switch. Go back downstairs and turn the power back on to the furnace or air handler. As the fan starts, turn the power off if the fan becomes unusually noisy (Fig. 8-19). The result should be the smooth sound of air flowing through the furnace.

If you have a heat pump thermostat that doesn't have a fan switch with an automatic and an on position, move the cooling setting down below room temperature, thus turning on the air conditioner. When you turn on the power again, the air conditioning function and the furnace fan should start. Run the thermostat in this mode only long enough to check the operation of the fans, then turn off the power switch of the furnace. Whenever you operate the air conditioning function or the heating function of the heat pump, *don't* restart the unit for about five minutes after shutdown. The refrigeration pressures then will have a chance to equalize within the piping. If the unit is restarted by mistake, you may blow the heat

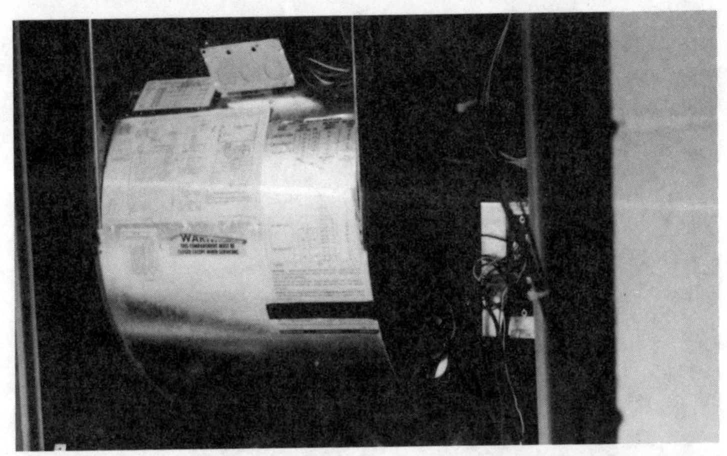

Fig. 8-19. The indoor fan assembly in an electric furnace. Note the restricted working space around the squirrel-cage fan. The fan motor is located within the fan.

pump fuse or circuit breaker. It's normal for the fuse or circuit breaker to blow. The compressor may not be able to start against a head pressure. Before resetting the circuit, you will still have to wait five minutes.

The indoor coil section of the heat pump system may never need cleaning if you take care of the air filter (Figs. 8-20A and 8-20B). If the indoor coil ices over and the air flow stops during air conditioning, the coil may be dirty. Check that the unit isn't low on refrigerant gas, and that the air filter is clean. Lack of gas or poor air flow will act just like a clogged coil. Call in a qualified serviceman. If you check the coil for contamination, you will probably have to cut an inspection hole into the ductwork. Imagine in which direction the air enters the coil, then cut an appropriately sized hole for the inspection. Never cut any hole in the casing of a furnace or air handler. The clogged coil condition will always be on the air-entering side of the coil or radiator. If you find that the inspection hole isn't possible without cutting a hole in the furnace, then the coil may have to be removed just for the inspection. Unless you're 90 percent sure of having a clogged coil, don't remove it for inspection. Removing the coil will cost you money and time. It's possible that the icing is caused from blocked or turned off registers.

TROUBLESHOOTING CHARTS

In Figs. 8-21 through 8-27 we have listed various causes of breakdown in electric, gas, and oil furnaces and heat pumps. Items

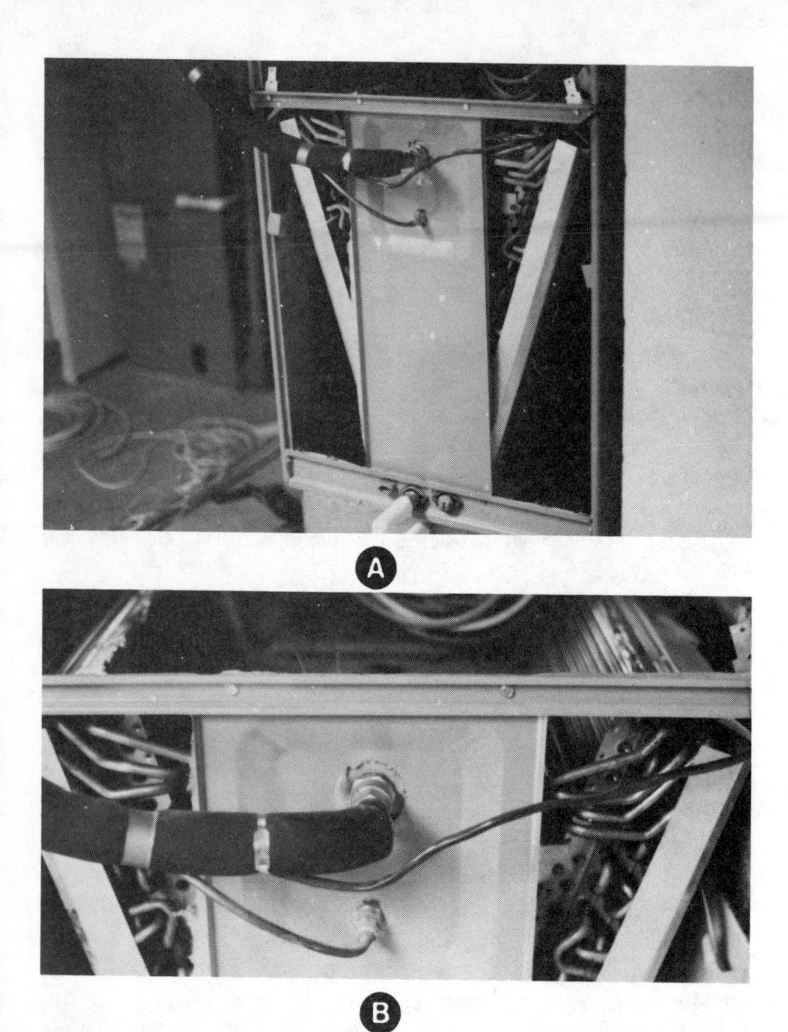

Fig. 8-20. (A) An inverted A coil installation mounted under an electric furnace. (B) The air flow is from the bottom up through the coil and into the electric furnace.

followed by an asterisk are the most common repair problems. If you find a problem, you must determine if the defect was caused by another part.

GENERAL CAUSES AND EFFECTS OF FURNACE FAILURES

A defective thermostat will cause the furnace to act as though heat isn't needed in the air conditioned space. Electrically the switch will be open, and it will remain so until it is repaired. To test

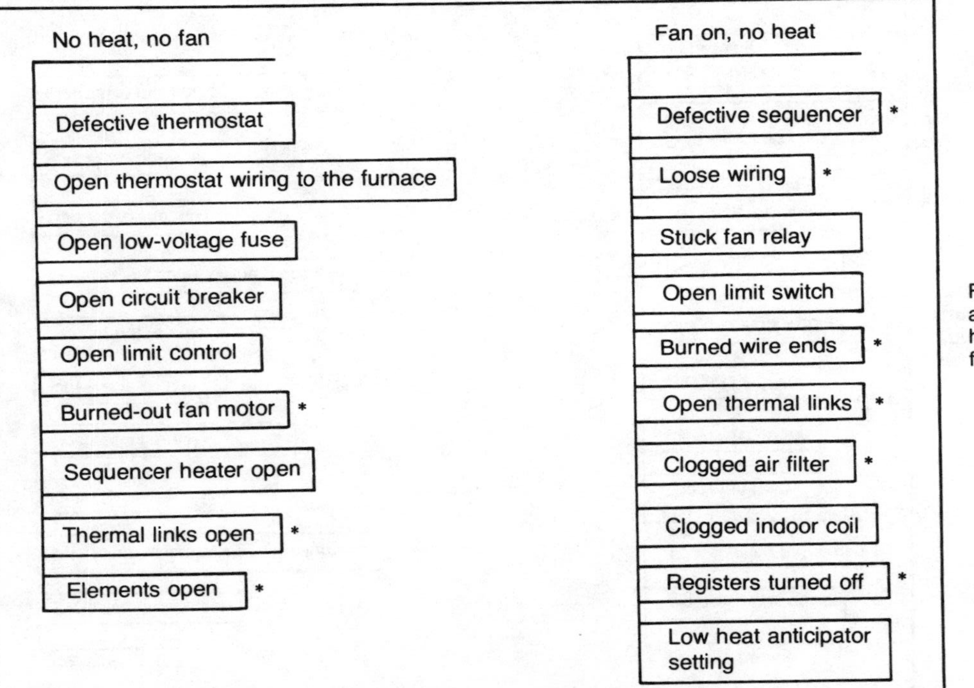

Fig. 8-21. Troubleshooting chart for an electric furnace—no heat. A low heat anticipator setting will cause the furnace to short cycle.

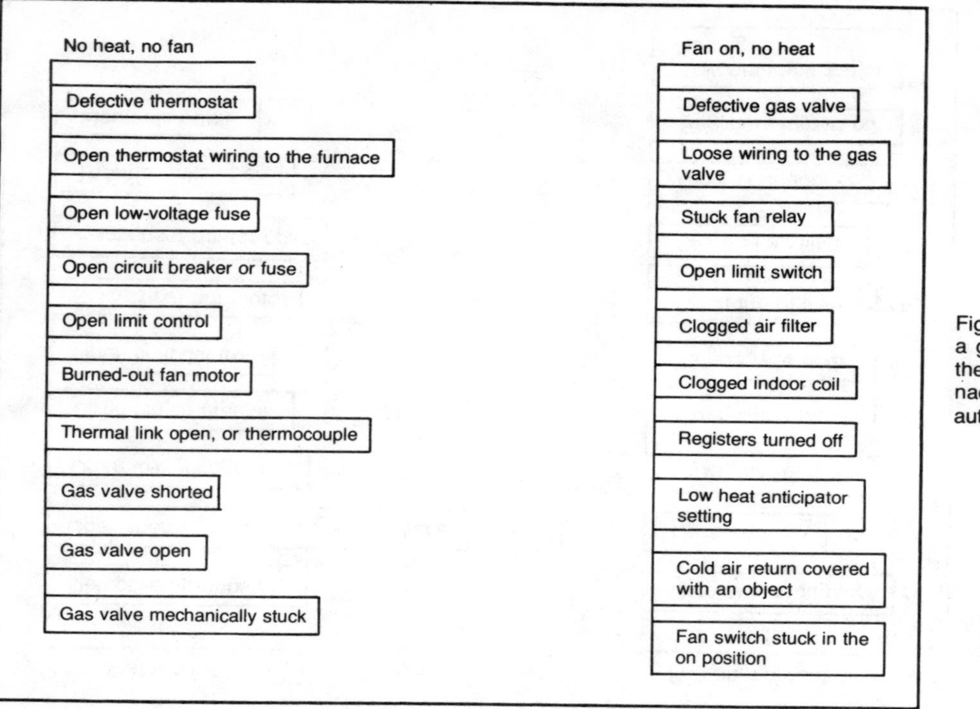

Fig. 8-22. Troubleshooting chart for a gas furnace—no heat. If most of the registers are turned off, the furnace will overheat and turn itself off automatically.

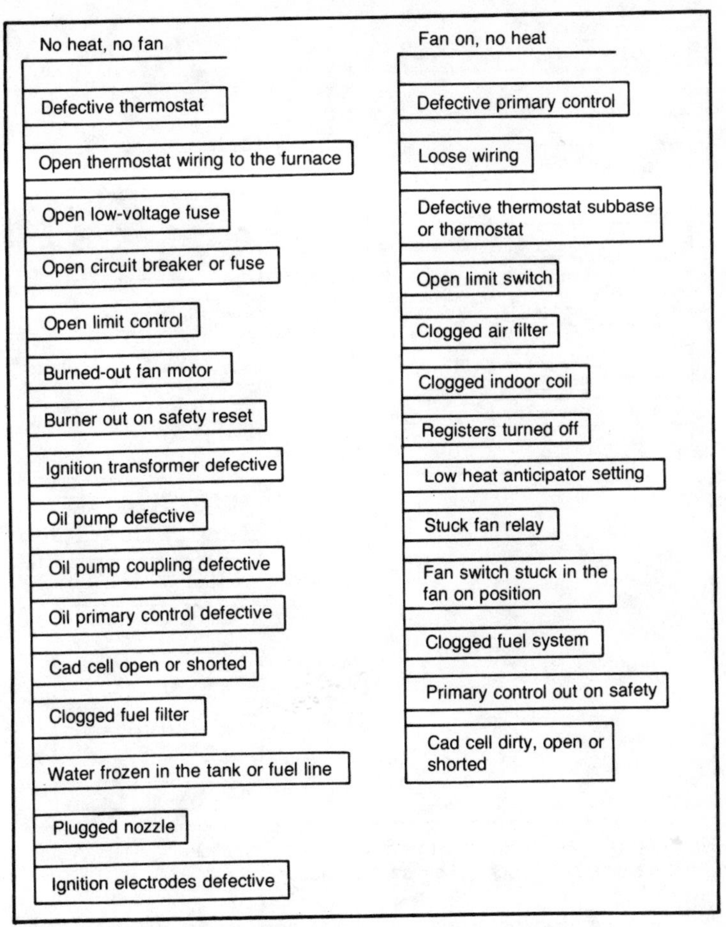

Fig. 8-23. Troubleshooting chart for an oil furnace—no heat.

for a defective thermostat, a voltage reading will be required at the furnace. Turn the thermostat up to its maximum value and apply the voltmeter test leads across the thermostat wiring terminals on the furnace. If you read a voltage that's the same as the control voltage, the thermostat is open, or the wire leading to the thermostat is electrically open. Control voltages in residential heating applications are limited to 24 volts ac 99 percent of the time. If you are in doubt, refer to the furnace wiring diagram. The other possibilities are line voltage (120 volts ac) and special direct current (dc) control schemes. A dc voltage will not register on the ac voltage scale of your meter. It's important that you know what should be present

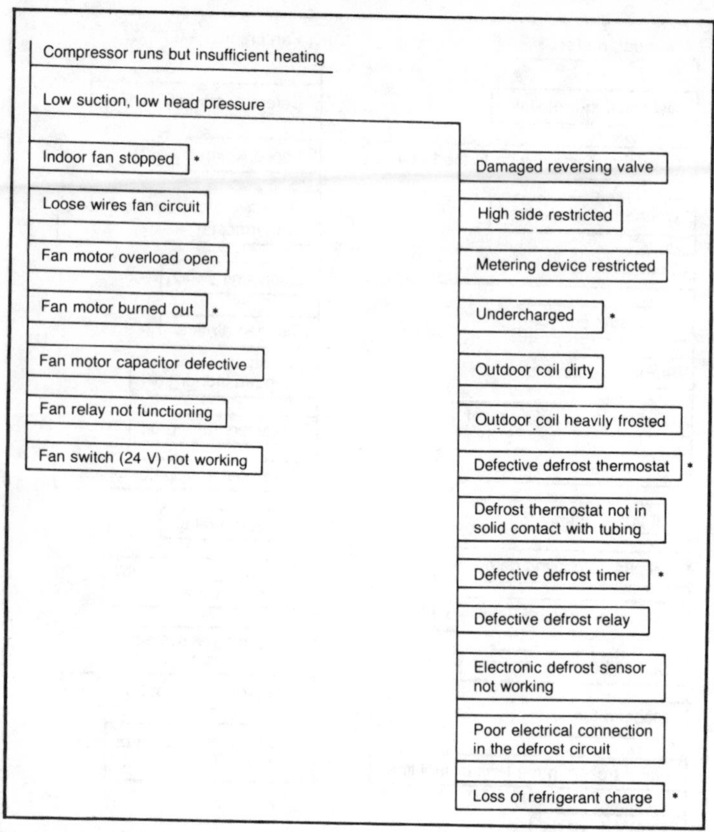

Fig. 8-24. Troubleshooting chart for a heat pump—insufficient heating. A defective defrost relay or clock will cause the outdoor coil to ice up.

before you measure the voltage in any electrical circuit. Thermostat problems are rare in gas or oil heating applications. The rate of incidence is somewhat higher in electric furnace applications because of the amount of control current drawn through the control circuit.

Open thermostat wiring is rare. If you determine that the wire is open through voltage measurements, a series of visual checks will be necessary. The most likely places for a wiring problem will be at wiring terminals or at a splice point in the wiring.

An open low-voltage fuse can be found by turning the room thermostat above the room temperature. Measure the voltage across the fuse holder terminals. If the measurement indicates full control voltage, the fuse is blown or open electrically. You can check the fuse with an ohmmeter.

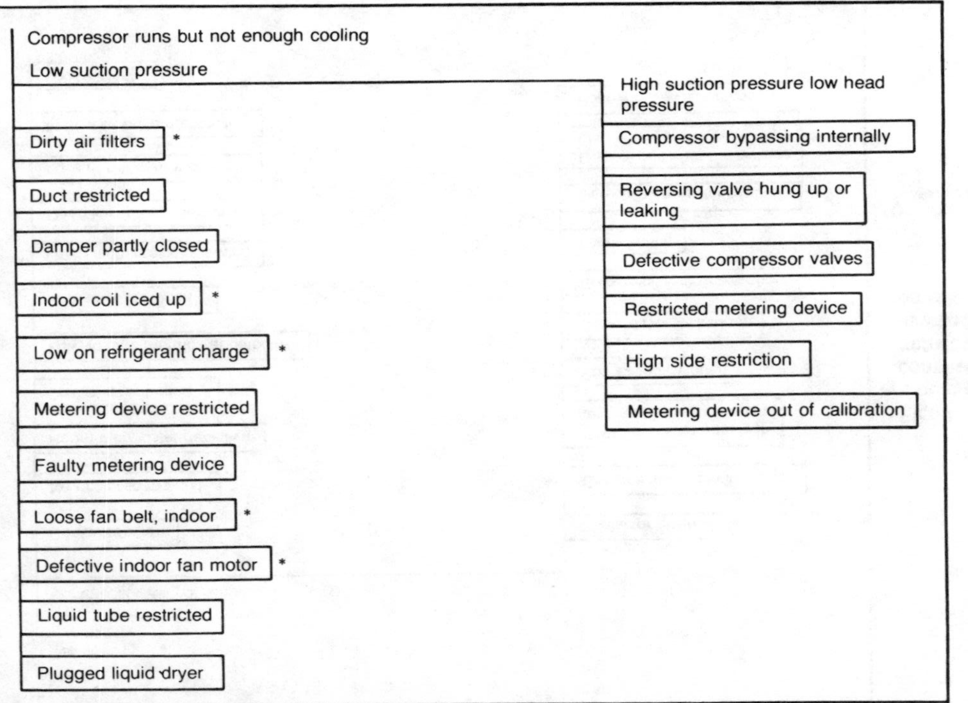

Fig. 8-25. Troubleshooting chart for a heat pump—insufficient cooling. If the indoor coil is iced up, it may be due to a dirty air filter or a restricted duct.

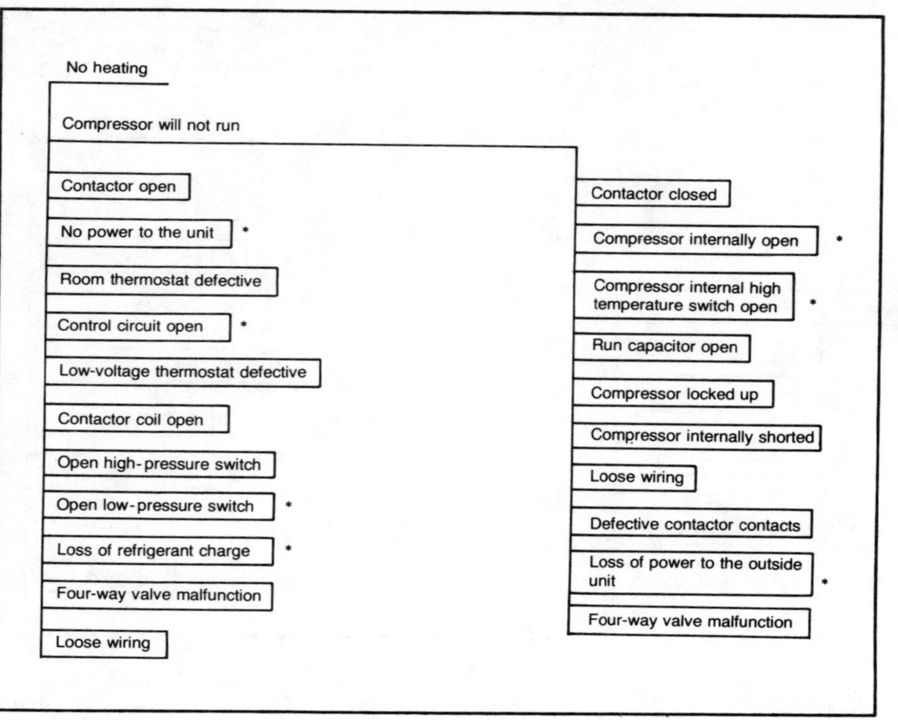

Fig. 8-26. Troubleshooting chart for a heat pump—no heat output. If the compressor is locked up, perhaps a defective low-pressure switch didn't function when the machine ran low on refrigerant.

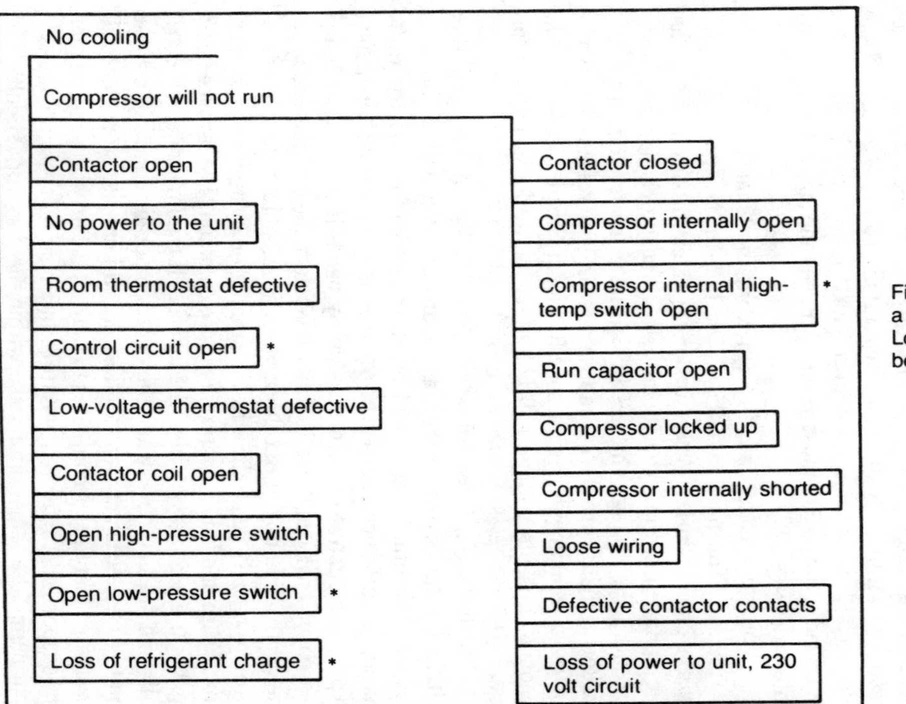

Fig. 8-27. Troubleshooting chart for a heat pump—no cooling output. Loss of the refrigerant charge may be a leaking fitting within the piping.

An open circuit breaker or fuse will give the effect of a completely dead furnace. Measure the voltage at the furnace after inspecting the circuit breaker or fuse. Depending on the electrical codes and manufacturer, you may have more than one set of fuses in the furnace circuit, but not more than two. Electric furnaces may have several wires feeding power to the heating unit. Gas and oil furnaces use only one feed circuit, but they may be double-fused in the fuse box for the house and possibly at the heating unit. Low supply voltage will be an indication of a blown fuse or circuit breaker in an electric furnace. No voltage at the furnace will indicate a blown fuse or circuit breaker in a gas or oil furnace. When checking fuses, the easiest method is to use an ohmmeter after the fuse is removed from the fuse holder. When in doubt about whether or not a circuit breaker has blown, turn the breaker off and then back on, thus resetting it internally. If a fuse or circuit breaker blows out or opens electrically more than once in a short period, the cause of the overload in the circuit must be found. Fuses and circuit breakers are safety devices.

Voltage measurements in heating units at times can be perplexing. You may have 110, 118, 120, or 125 volts ac at the average house outlet. When working with circuits, remember that the voltage entering your home is continually changing to some extent. If the voltage changes by 3 volts or more, you will notice flickering lights. In most cases a variance of 10 percent is acceptable for most equipment design specifications. Small changes in the amount of usable voltage will not affect the performance of a furnace.

An open limit control will be caused by high temperature within the furnace or a mechanical malfunction within the limit control. The limit control will break the flow of electricity to the burner if the duct temperature exceeds a preset value. Never bypass the control or turn the dial by hand. If the control is bypassed, you will run the risk of fire. If you turn the dial plate by hand, the control will go out of calibration. Causes of high duct temperature are usually related to the air circulation system or lack of air flow. Filter and fan motor problems are common. The furnace heats up, but the heat stays within the furnace. If the limit opens electrically and doesn't close when the duct temperature is reduced, the limit has a mechanical malfunction and will have to be replaced. If you notice that the fan continues to run, but the fire turns on and then off, suspect duct over temperature. In a heating system without a heat pump, the temperature rise within the ductwork shouldn't be more than 50° F. Temperature rise is the air going into the furnace compared to the air temperature coming out of the furnace. Check for poor air flow if the temperature rise is more than 50° F.

A burned-out fan motor will cause the furnace to overheat because of a lack of air flow. The fan motor bearings can freeze up, the winding can burn out, or a simple open circuit can occur within the motor. With the power turned off, smell the motor. If it doesn't smell burned inside, it's probably repairable. If the motor is very hot to your touch, frozen bearings or a run capacitor is probably the cause. If your fan is a belt-drive unit and the motor is hot, a starting switch is the probable cause of failure if the bearings are free. It usually will be less expensive to have a motor repaired than to buy a new motor. The people at the motor shop can estimate the repair cost and advise you.

Links

An open thermal link or thermocouple is probably the most common cause of gas furnace failure. As long as the thermocouple detects the heat of the pilot light, it will generate 30, 500, or 750 millivolts (one millivolt = 1/1000 volt). If the heat of the pilot is lost or if breakdown occurs within the thermocouple, the gas furnace will automatically turn off, stopping all gas flow. The thermocouple is a safety device and must operate properly. Follow the relighting procedure listed on your furnace if you determine that the cause of an outage is a pilot flame that's gone out. If the pilot fails to stay lighted after the pilot flame has been established for 90 seconds or if the pilot fails continually, you should suspect that the thermocouple or gas valve is faulty. Nine times out of 10 it's the thermocouple that is at fault. Because of the costs involved, change the thermocouple first and test it for a week or less. Most gas furnace failures are due to a defective thermocouple. When a thermocouple is installed, the connection to the main gas valve should not be tightened more than one-quarter turn, or damage may result within the gas valve. Call a qualified gas furnace repairman.

A fusible link opens electrically if the temperature in a detected area exceeds the rating of the fusible link. A fusible link looks like a metal-covered diode and is wired directly into a controlling circuit. As heat around the thermal link increases past the critical point, a low-temperature metal melts within the device and opens an electrical circuit that the metal had bridged. When the link has not been opened, it will not reset itself. When the link is open electrically, it must be replaced to effect a repair. Fusible links are placed at critical points in electric and gas furnaces. If a link opens in either type, you must determine why. The links are somewhat expensive and serve a definite purpose. Too often fusible links in the field are bridged or removed from the circuits that they are protecting.

Valves

A shorted gas valve will always cause the thermostat or possibly the low-voltage transformer to burn out. A few furnaces will have a low-voltage fuse that will blow in a shorted gas valve situation. Shorted main gas valves should be suspect if a thermostat or transformer fails. The amount of control current is stamped on most gas valves by the manufacturer. The control current usually ranges from 0.2 to 0.65 amp. If by measurement the current is somewhat higher than the stamped rating, the gas valve must be replaced. If a main gas valve is defective, it must be replaced by a qualified gas serviceman.

An electrically open gas valve will not turn on the main gas flow, no matter where the thermostat is set above room temperature. To test the gas valve, measure the voltage across the terminals of the gas valve coil. If the valve is turned on mechanically and the main fire doesn't light, the valve is defective. The pilot must appear to function normally for the voltage test to be valid. You also can test the coil with an ohmmeter. The resistance of the coil depends on the amount of voltage used as control actuation. A completely open gas valve will register infinity on the ohmmeter scale. A qualified gas serviceman must replace the gas valve.

A mechanically stuck gas valve is somewhat rare. A gas valve can stick when it's both open and closed, so the result may be no heat at all or too much heat. If you find that the gas valve is warm and that you have voltage across the coil terminals, but there isn't a fire, the main gas valve is mechanically stuck if the pilot is burning. Conversely, if the valve body is cool to the touch and no voltage is measured across the gas valve coil, but the main fire refuses to extinguish, the main gas valve is mechanically stuck open. Gas valves used on residential furnaces generally are not repairable. It's better to install a new unit, since the valve is a primary safety control. Replacement of the gas-handling parts of a furnace should only be done by a qualified gas serviceman.

So far we've considered heat source problems. If the fire fails, the furnace won't generate heat in the ductwork of a gas furnace. If heat isn't present in the ductwork, the forced air fan sensor won't start the ducted fan in the majority of gas-fired residential furnaces. The result is no heat and no fan. If you have a good background in electricity, you will be able to repair about 50 percent of the problems encountered in gas or electric furnaces. When the task is too difficult, call a repairman.

Heating Elements

Elements that are open electrically are a fault found only in electric furnaces. Most electric furnaces will have between two and six heating elements that generate the heat needed for the home. If any one or several of the elements open electrically, the result is diminished heating capacity. Each heating element in a furnace has the ability to turn electrical power or amps into Btu of heat. The complaint that we hear most often when a burned-out element is found is that the furnace works, but there just isn't enough heat. You probably won't even know that you have a defective element until the coldest day of the year. Most furnace installations have more than enough capacity except for the day that the outside temperature is −10° F. and falling. If you suspect that an element isn't doing its job, turn the thermostat at least 10° above the room temperature. Wait about 10 minutes and measure the air temperature going into the cold air return. Also, measure the air temperature leaving one of the supply registers. Remove all power from the furnace and remove the access panels. Turn the power back on and wait another 10 minutes.

Using an ammeter, measure the current flowing through the wires that supply power to each element. The amount of current flow will be between 15 and 20 amps per element. If you find that you have no current flow or abnormally low current flow, suspect the element or fusible line to be faulty. It's also possible to test elements with an ohmmeter, but this is more time-consuming. When testing with an ohmmeter, the furnace remains off. When working on a piece of electrical equipment, it's a good idea to tape the power switch or circuit breakers in the off position. This is done to preclude the possibility that the circuit will be energized while you're working (Fig. 8-28).

Sequencers

Electric furnace sequencers and/or relays are used to turn on the electric heating elements from the command of the thermostat. Thermostat circuits are generally 24-volt circuits. The heating element circuit operates from 208 to 230 volts ac. In a relay-

Fig. 8-28. When you are working on an electric furnace, turn the electricity off before removing any access panels. More than one switch may have to be turned off. The furnace may have more than one supply circuit.

operated electric furnace the thermostat will call for heat, which will cause all the heating element relays to close simultaneously. The instant turn-on feature of the relay system may cause power line problems because of the sudden rush of current into the furnace. Sequencers fulfill the same function as the relay system, but the difference is that they turn the elements on one at a time. Power companies prefer sequencers because they reduce the possibility that the neighborhood lights won't flicker each time your furnace turns on.

Sequencers and relay systems are very dependable, but they occasionally fail. If you find that an element isn't drawing power, check the voltage across the element terminals. If you don't read full supply voltage across the element terminals, you may have a defective sequencer or relay for that element. The voltage test across the elements has to be done with the furnace energized and calling for heat. Sequencer systems are sometimes slow. Wait about 10 minutes before testing to allow enough time for the sequencers to function. The waiting period isn't required for relay systems.

If you are testing a sequencer furnace, refer to the wiring diagram. If multiple sequencers are used in your furnace (some furnaces may only use one sequencer), you will note on the diagram that the sequencers cause each other to turn on in sequence. Sequencer number one turns on sequencer number two, number two turns on number three, number three turns on four, etc. Sequencers handle from one element to four elements each. The number of sequencers is different for different manufacturers.

The sequencers in use operate from 24 volts and switch 208- to 230-volt circuits. If the wires are transposed from the 230 side to the 24, the thermostat may be blown off the wall. When changing a sequencer, be absolutely sure that you know the function of each wire. The wires must not be applied incorrectly at any of the wiring terminals of that sequencer, or a hazard will result. If your furnace uses a thermostat wired with small thermostat wire (usually 18-gauge), never use more than 24 volts ac in that wire. Follow your local code or the national electrical codes. When you test any electrical circuit, use a combination of voltage and current testing. The current test will tell you if you have a closed circuit. The voltage test will tell you what switches, wire, element, or limit switch is open. Find a circuit on the unit wiring diagram. Visualize how the circuit works, then begin testing that particular circuit. The wiring diagram will tell you what the normal condition should be. Read the meter and decide if you have a normal or abnormal condition.

Burned Wire Ends

Burned wire ends in electric furnaces are a common problem. If a power wire has a spade connector that becomes loose, it will overheat and burn off the connector. The primary trouble spots are at any points that a power wire can become loose. When a wire carrying a lot of current is loose, it will heat up. The hotter it gets, generally the looser it gets. The result is that it will burn in half. With the furnace turned off, an inspection can be made of the wiring. If you find a wire that is burned off, repair it with a new spade connector. Don't use a section of the old wire that has been overheated and discolored. Cut the wire back to a point where the copper is bright and shiny. Use a good crimping tool. The wires in any furnace are of the special high-temperature variety.

Loose Wiring

Loose wiring in gas, electric, or oil furnaces is rare in the field. With the heating unit turned off, a check for tight screws takes little time. Quality control seems to be quite good in the factory as far as application of wiring is concerned. A wire nut connection is more prone to failure. Only one type of heating system is susceptible to even a slightly loose wiring connection. That heating system is a millivolt-operated gas system. Oxide on one of the interconnecting wires is enough to stop the very small amount of current that the circuit needs for operation.

Stuck Fan Relay

Stuck fan relay problems occur in older furnaces to some extent because of the contacts within the relay welding together. The contacts are coated with a hard surface that retards the welding action because of the continual arcing experienced by the contacts during normal operation. The result of welded fan relay contacts is that the fan won't turn off. It runs continuously, no matter where the thermostat is set. The only solution is to replace the relay. Contact filing is useless as a repair procedure.

Clogged Air Filter

A clogged air filter will cause a furnace to overheat, even though the fan is running. The furnace requires a specific volume of air for internal cooling. When that air is reduced, the furnace will overheat. When the furnace overheats, it will automatically turn itself off. Extremely high temperatures are possible within the home and furnace if the high-temperature switch fails to function. To test a high-temperature cutout switch, stop the air flow and wait a reasonable amount of time. The furnace should turn itself off within 10 minutes or less. Never remove a high-temperature switch or limit control from your furnace.

Registers

Registers that are turned off are another cause of furnaces not operating properly. The furnace will overheat. Air flow must be maintained for normal furnace operation. The temperature of the air leaving the furnace should gain no more than 50° F. when compared with the air entering the furnace.

Heat Anticipator

A heat anticipator is contained within most modern thermostats. It artificially heats the interior of the remote thermostat so that it turns the furnace off, just before room temperature is attained. By using a heat anticipator in a thermostat, better control is obtained. Control within 1° isn't uncommon if the anticipator is set properly. The temperature control is only at the area of the thermostat—not in a room 50′ away. If a thermostat is nonanticipated, the room thermostat won't turn off until the air temperature is at the setting of the thermostat. If the thermostat is set at 70° and the furnace turns off at 70°, you still have a hot furnace to cool. As the rest of the heat is blown from the furnace, the room temperature rises above the thermostat setting. This will be experienced in every heating cycle with a nonanticipated thermostat used as a control.

The best way to set a thermostat heat anticipator is by direct measurement. Using an ac ammeter in series with the thermostat, measure the current flow when the thermostat is turned above room temperature. The ammeter should have a scale with the capability to read currents from 0.2 amp to 1.5 amps. As you turn up the thermostat, watch the dial of the ammeter and make note of the reading attained. Gas furnaces will give a reading of 0.2 to 0.65 amp generally, while oil furnaces will normally run from 0.4 to 0.60 amp. If the reading varies somewhat from these examples, you need to establish the amount of current that flows through the thermostat circuit.

Electric furnaces have thermostat currents that are higher than those of gas or oil furnaces. Sequencers or relays draw more current than the average gas or oil furnace because of the complexity of the control system. An electric furnace will usually draw from 0.6 to 1.2 amps of control current. Sometimes when testing the thermostat current draw in electric furnaces, the thermostat current won't stabilize for some minutes after the command is given to start the heating cycle. Regarding gas and oil furnaces, the heat anticipator dial inside the thermostat will be set to the same value read on the ammeter. Let's say your furnace draws 0.5 amp on the ammeter test, then determine where 0.5 amp is located on the heat anticipation scale inside the thermostat and set the pointer to 0.5 amp. That's all there is to it on gas and oil furnaces.

The procedure is the same when setting an electric furnace anticipator, as long as the current draw is steady through the thermostat circuit. If you find the current value unstable due to the sequencer coming on line, the only recourse is to wait for a maximum reading on the ammeter and set the anticipator pointer to that value. If the current surges, then falls back to a lower number on the scale, the pointer setting will have to be averaged. If the room gets cold before the furnace turns back on, the heat anticipator is probably set too high. If the thermostat has to be set to a higher than normal temperature to maintain a reasonable level of comfort, the setting is probably too low.

Modern thermostats, those with glass tubes that contain mercury, must be mounted level to be even nearly accurate. Leveling lines or pegs are provided by the manufacturers and are usually on the inside of the thermostat. Before any anticipator adjustments are made, the thermostat will have to be level. You can set a thermostat heat anticipator by trial and error, but it's time-consuming. If the room temperature is overshooting, lower the pointer setting in small segments until the temperature range evens out. If the temperature is undershooting, increase the pointer setting in small segments until reasonable temperatures are attained. Remember that being cold before the furnace turns on means that the setting is too high. If you never get up to temperature, the anticipator is probably set too low. Remember that the thermostat must be level before starting anticipator adjustments. Heat anticipators are used only on gas, oil, or electric furnaces. Electric baseboard heating systems don't use anticipation in line-voltage thermostats, nor do millivolt gas systems.

Oil-Fired Furnaces

Oil must be pumped into oil-fired furnaces under pressure. The oil has to be ignited, then the proper amount of air has to be mixed with the oil spray. As the oil heats the furnace, a method must be used to automatically turn on the forced air fan at the desired temperature. A safety device must be used that stops the oil supply in case of flame failure.

The efficiency of an oil-fired furnace is directly related to the oil-air mixture ratio and the design of the heat exchanger. If a furnace has a poor grade of fire or a poorly designed heat exchanger, it's going to cost you heating dollars. If you burn a certain quantity of fuel per hour and mix that fuel with excessive air for combustion, there will be an efficiency loss. Excessive air will act like a fluid and carry more heat out of the chimney, or it will cause incomplete burning of the oil spray. Either way, the cost of operation is increased. If the same quantity of oil to the furnace is delivered, but at a less than needed quantity of air, soot will result. The soot acts as an insulator, causing a reduction in efficiency and an increase in operating costs. Oil burner nozzles are a prime cause of oil furnace failure. They become clogged or coated with contamination that distorts the oil spray. When the spray is distorted, the air-to-oil mixture ratio is poor, causing added operating cost or a complete failure of the system.

Routine maintenance and efficiency testing of an oil furnace is extremely important for economy. Special instruments are required to test any high-pressure oil burner. Notice what the fire looks like after the furnace is serviced and periodically check the appearance of the fire. If the color, shape, or sound of the fire changes, the burner will require service. Oil nozzles aren't cleanable unless special precautions are taken; they should just be changed when needed.

The amount of chimney draft (shouldn't exceed 0.04″ water column vacuum) or excessive supply duct temperature are directly related to efficiency. As the chimney vacuum increases, it pulls the heat out of the furnace at a rate faster than what's needed for the heat to be absorbed into the heat exchanger. If the furnace fan doesn't move enough air, the supply duct temperature will operate in excess of the design limits of the furnace. A loss of efficiency will occur. The air temperature leaving the furnace shouldn't be higher than 50° F. over the air temperature entering the furnace through the return duct. If the air entering the furnace is measured at 68° F., the air leaving the furnace shouldn't be higher than 118° F. for maximum efficiency.

An oil furnace also contains an ignition transformer rated at 10,000 volts ac at 0.025 amp. If the transformer voltage drops 10 percent, ignition problems may become apparent within the heating system. When this much electricity is used in any circuit, cleanliness is important. Dirt, lint, or soot will act as a wire or conducting path and bleed away part or all of the ignition voltage. Delayed ignition is caused by ignition faults.

The oil pump is either single- or two-stage pumping, with an output pressure of at least 100 psi. All nozzles are rated at 100 psi of pressure. The pressure fed to the nozzle from the pump should not be less than the rating pressure. If the pressure is less than 100 psi, distortion will occur in the fire, plus unstable operating conditions and reduced efficiency.

A *burner motor* drives the power burner fan and oil pump. The fan is directly connected to the burner motor. The pump is interconnected by a rubber coupling meant to compensate for slight misalignment of the motor and pump. Broken couplings are common in the field.

The main burner control on an oil furnace is called a primary control or *stack switch.* The thermostat gives the command to supply the living space with heat. The command is received by the primary control, which then activates the power burner. The primary control also includes a fire sensor that turns off the burner if the fire fails to light or should fail during heating operation. The safety function of the primary control shouldn't be reset more than three times because the inside of the furnace may become loaded with raw oil.

The forced air fan section of oil furnaces is the same as that of a gas-fired unit. The fan is activated by a temperature-sensitive switch and deactivated in the same manner.

HIGH-PRESSURE CUTOUTS

The purpose of a high-pressure cutout in a refrigeration unit is to turn off the compressor if a high head pressure should result from a mechanical failure. If an excessive high pressure was to be created within the refrigerant piping, the piping itself or the compressor casing could rupture and cause damage or injury.

A high-pressure cutout should never be bypassed in such a way that would cause its failure in an emergency. We've found high-pressure switches jumpered inside outdoor units.

The automatic reset high-pressure cutout will open a switch between 350 to 450 pounds per square inch gauge (psig). When the cutout switch opens, it will automatically reset at about 250 psig. Cut-in and cutout pressures vary among the various models of safety switches. The equipment in which the safety switch is used will determine the pressure setting required for the safety switch. Note the difference between the cutout pressure and the cut-in pressure. The difference is called the dead band or switchover pressure. The wide band of pressure difference is built into the switch to prevent short cycling. The manual reset high-pressure cutout is the kind in which the control must be manually reset before the equipment will restart the compressor. The manual reset type will open a switch and cause the compressor to stop, just as in the automatic reset type. The pressure that a manual reset control will trip is the same as the automatic control, but it will not reset automatically. When either the automatic or manual cutouts function, they should be within plus or minus 20 percent of their rated pressure for opening or closing. If you find a switch that doesn't function within the 20 percent limit, the safety switch should be replaced.

Refrigeration servicemen use special gauges and gauge hoses that are rated at 500 psig. Never use nonrated gauges or hoses when checking refrigeration systems. Excessive pressure may cause a nonrated gauge hose combination to rupture.

If a high-pressure switch should open, it is identical to an open household light switch. It may use different amounts of voltage in its application. If you measure across the two wires leading to the high-pressure switch with a voltmeter and read the voltage in the circuit that contains the switch, it is open. If no voltage reading is obtained when you measure across the switch (and the circuit is energized), the switch is closed electrically. Another method of checking switches is to remove power from the switch, disconnect the wires from the switch, and use a continuity tester or ohmmeter.

High-pressure cutout switches are manufactured in several configurations. The best way to locate a switch is to use the wiring diagram for the equipment being checked and follow the wires until the device is located. The smaller high- and low-pressure switches are somewhat fragile. The wire terminals will break off if the switch is bumped.

Never let the heat pressure of an air conditioner, heat pump, or refrigeration unit exceed 450 psig. Excessive pressure may cause the piping or gauge set to blow apart.

The usual causes for high head pressure in excess of the normal operational limits of a machine are as follows:

Dirty Condensing Coil. The condensing coil in an air conditioner will be part of the outside unit. The condensing coil in a heat pump will switch from the outside unit to the inside furnace or air handler. If in the air conditioning function, the condensing coil will always be located in the outdoor section of either a heat pump or air conditioner.

Defective Condensing Fan Motor. If the fan motor should become inoperative for any reason, the head pressure will increase very rapidly. If the compressor is running, the condensing fan motor must also function. Exceptions exist, but usually not in the smaller equipment.

Defective Condensing Fan Motor Relay. A relay is used as a switching mechanism to turn the motor on and off as needed. Some units will use the same relay that the compressor uses for condenser fan power.

Condenser Fan and Coil Air Recirculating. If the air leaving the condenser coil is confined and forced to recirculate back through the condensing coil, the head pressure will increase because of the increased air temperature passing through the coil.

Condenser Coil Air Temperature in Excess of 115° F.

Fig. 8-29. The outdoor section of a heat pump will contain a fan. The outdoor fan is responsible for all the air movement in the outdoor unit. The fan blade is attached to the drive shaft of the motor in most outdoor units.

(Fig. 8-29). Design will limit the outside air temperature that the condensing coil can use for the condensing function of the refrigeration cycle. If the outside air temperature should exceed the design limit, excessive head pressure will result. If the outdoor section is located in a hot area, the air temperature may exceed the design limit.

FORMULAS FOR CALCULATING AIR FLOW

- Cubic feet of air per minute = CFM.
- Kilowatt = 1,000 watts,
- ΔT = delta temperature, or the difference in two temperature measurements.
- 1.08 = air flow constant, specific heat.
- 3413 = number of Btu in 1,000 watts of electrical power.
- $\text{CFM} = \dfrac{\text{kW} \times 3413}{\Delta T \times 1.08}$
- $\text{Btu} = \text{CFM} \times \Delta T \times 1.08$.

If measurement of air flow in an electric furnace is desired, the kilowatts may be found on the specification plate. The air flow quantity may be found using the above calculation. If you want to determine the total air flow of a gas or oil furnace, use the following:

Btu output of the furnace = 0.75 × Btu of input fire (oil)
Btu output of the furnace = 0.80 × Btu of input fire (gas)

1 gallon of oil burned in one hour = 140,000 Btu

Average furnace efficiency for oil = 75 percent (efficiency = output/input)
Average furnace efficiency for gas = 80 percent (efficiency = output/input)

Btu input may be found on an oil furnace by knowing what size nozzle is installed in the burner assembly and by knowing that the pump pressure is 100 psi. (Nozzles are rated at 100 psi). If the nozzle was printed with .75, the nozzle delivers 0.75 gallon of oil per hour at 100 psi. If you inject 0.75 gallon of oil into the furnace, the amount of oil input into the furnace equals 0.75 × 140,000 (140,000 Btu = 1 gallon of oil) = 105,000 Btu input. If you want to find the output value in Btu of the oil furnace, then compute the following:

Btu output = Btu input × 75 percent efficiency

In our example, the oil furnace with 105,000 Btu input equals 105,000 × 0.75 (percent) = 78,750 Btu into the duct system. A 25 percent loss is assumed because of chimney heat. Newer furnaces may be 80 percent efficient.

$$\text{CFM} = \frac{\text{Btu output}}{\Delta T \times 1.08} \quad \text{true for gas or oil furnaces}$$

Temperature difference (delta T) may be measured in the furnace ductwork. The air-entering temperature may be measured at the air filter. The air-leaving temperature should be measured 10′ away from the furnace in a supply duct. Run the furnace 10 minutes or more to stabilize duct temperatures. Delta T equals supply temperature minus the return air temperature. The supply air is warm air delivered to the home living space. The return air is air returning to the furnace from the living space.

Appendix

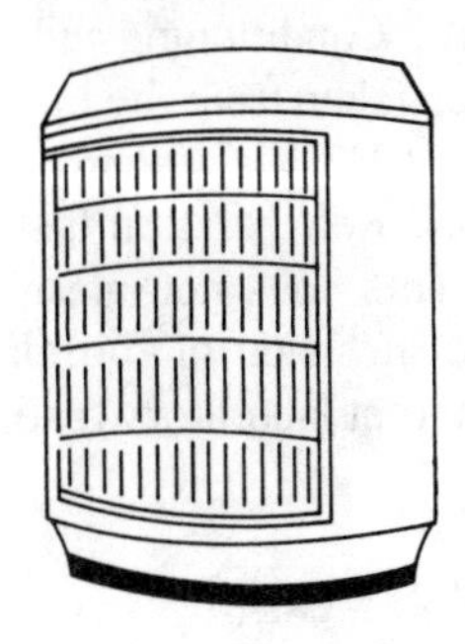

Heat Pump Efficiency Ratings

Testing and evaluation of air conditioners and heat pumps is the function of the Air Conditioning and Refrigeration Institute (ARI). The Institute also tests solar collectors and sound rates outdoor unitary equipment.

Because of space limitations, only a few of the many heat pump manufacturers are represented here. The manufacturers were chosen for exceptional equipment or for being known by the general public.

If you would like a complete ARI rating book, send $3 to Air Conditioning and Refrigeration Institute, 1815 North Fort Myer Drive, Arlington, VA 22209. The ARI publishes a revised directory twice a year. We highly recommend that you use these ratings when selecting a heat pump for your home or business. The listings will enable you to choose an efficient heat pump. Compare and you will achieve a maximum dollar return on your investment.

All manufacturers' model numbers are listed with various combinations of outside and inside units. If you are given an estimate for an installation, you'll have the ability to compare unit efficiencies before you buy. If your primary use of the heat pump is in the air conditioning mode, look for the highest EER (energy efficient ratio) or SEER (seasonal energy efficiency ratio) rating to obtain maximum savings. If heating is your main concern, look for the highest COP (coefficient of performance) rating in the Btuh range that you require. The larger the EER, COP, SEER, or HSPF (heating seasonal performance factor), the better the unit at least as far as efficiency is concerned.

The following heat pump efficiency information has been provided courtesy of the Air Conditioning and Refrigeration Institute. All information has been taken from the institute's directory of July 1 through December 31, 1982. We wish to convey a special thanks to the Air Conditioning and Refrigeration Institute for permission to reprint this material. This does not mean that the Institute endorses this book. The inclusion of the efficiency information is meant to provide competent laboratory testing data for prospective heat pump purchasers.

WHAT ARI CERTIFICATION MEANS

This directory of certified air source unitary heat pumps lists all eligible models produced by selected manufacturers participating in the applicable certification program of the Air Conditioning and Refrigeration Institute. Listing in the directory means that the models have been certified to ARI under the applicable standard to meet the ratings claimed for them by their producers and to perform under the test conditions described in the standards.

Testing procedures require that approximately 30 percent of each participating manufacturer's basic models undergo laboratory testing for rating and performance each test year. Participating manufacturers must file certification data with ARI on all models produced with the scope of the programs. These data are carefully evaluated by ARI engineers before models are listed in the directory, and tests are required of any models whose data appear questionable.

In addition to evaluation of the certified data and to the random testing of approximately 30 percent of all basic models each year, participating manufacturers may request that tests be made of competitors' models where the certified rating is questioned by them. If a model so tested fails to meet the rating claimed within the tolerances permitted or fails to meet the performance requirements, its manufacturer must pay the cost of the test. If it is found to perform within the limits required for certification, the manufacturer who challenged the model must pay the costs.

The manufacturer of a model that fails to pass the specified tests has two basic alternatives. He may rerate the model in question to reflect its tested capacity, or he may withdraw the model from his line. If one of these alternatives is not accomplished, the manufacturers' right to use the ARI certification symbol on all his models is withdrawn, and the manufacturer's name and listing is deleted from the directory.

Penalty for Misrepresentation of Performance. If it is demonstrated that a manufacturer has misrepresented the rating or other performance requirements of the ARI standard and has failed to make proper correction, the right to use the symbol will be withdrawn—not only for the specific model concerned, but for all models produced by that manufacturer. The manufacturer's name and listing will be removed from the directory.

Energy Efficiency Ratio (EER). A ratio calculated by dividing the cooling capacity in Btuh by the power input in watts at any given set of rating conditions. It is expressed in Btuh per watt (Btuh/watt).

Coefficient of Performance (COP). A ratio calculated by dividing the total heating capacity provided by the refrigeration system, including circulating fan heat but excluding supplementary resistance heat, (Btuh) by the total electrical input (watts) × 3.412.

Seasonal Energy Efficiency Ratio (SEER). The total cooling of a central air conditioner in Btu during its normal usage period for cooling (not to exceed 12 months) divided by the total electric energy input in watt-hours during the same period.

Heating Seasonal Performance Factor (HSPF). The total heating output of a heat pump during its normal annual usage period for heating divided by the total electric power input during the same period.

ARI RATING FOOTNOTES

§—Voluntarily revised since the last directory.

"WAS"—Indicates a rating that has been involuntarily changed since the last directory.

*—Highest sales volume tested combination required by DOE (Department of Energy) test procedure.

(1)—For application ratings, such as 200-volt or 208-volt ratings, refer to manufacturers' specifications, literature, and operating instructions.

(2)—For special application information, refer to manufacturers' specifications, literature, and operating instructions.

(3)—This model designation may specify an amount of electric heat in kilowatts. Refer to manufacturers' specifications, literature, and operating instructions.

(4)—For two-stage units and/or multiple door outdoor coil blower unit combinations, refer to manufacturer's specifications literature, and operating instructions.

HSP-A	single package heat pump, air source
HOSP-A	single package heat pump, air source, heating only
HRC-A-CB	heat pump with a remote outdoor coil
HORC-A-CB	heat pump with a remote outdoor coil, heating only
HRC-A-C	heat pump with a remote outdoor coil with no indoor fan
HORC-A-C	heat pump with a remote outdoor coil with no indoor fan, heating only
HRCU-A-CB	split system: heat pump with a remote outdoor unit, air source
HORCU-A-CB	split system: heat pump with a remote outdoor unit, air source, heating only
HRCU-A-C	split system: heat pump with a remote outdoor unit with no indoor fan, air source

AIRTEMP CORPORATION DOE Covered Products

Type: HSP-A Trade Name: Airtemp

footnote(s)	Model: outdoor unit	Model: indoor unit	Cooling capacity (Btuh)	SEER	Heating capacity (Btuh)	COP @ 47°	Heating capacity (Btuh)	COP @ 17°	HSPF
1	31020CQ111		22,400	7.60	22,000	2.50	12,000	1.70	5.35
	31030FK111		27,000	5.60	27,000	1.90	15,500	1.30	4.20
	31025CK111		28,000	7.50	31,400	2.60	18,000	1.80	5.90
1	31036FQ111		·32,000	6.40	32,000	1.80	18,500	1.30	4.20
1	31030CQ111		35,400	7.50	38,000	2.70	18,500	1.60	5.35
	31035CK111		40,000	6.50	47,000	2.30	26,400	1.60	5.35
	31040DK111		46,000	7.50	48,000	2.50	27,400	1.60	5.35
	31050DK111		57,000	7.50	60,000	2.50	35,000	1.70	5.60

Type: HRCU-A-CB Trade Name: Airtemp

footnote(s)	outdoor unit	indoor unit	Cooling capacity (Btuh)	SEER	Heating capacity (Btuh)	COP @ 47°	Heating capacity (Btuh)	COP @ 17°	HSPF
1	32918EQ111	BWRE-018-S7BA,DA	17,000	8.20	17,000	2.80	10,500	1.80	6.10
1	32918EQ111	BWRG-018-S3BA	17,000	8.20	17,500	2.80	10,500	1.80	6.10
1	32918EQ111	* BWRJ-024-C2AA	17,500	8.40	17,500	2.90	10,500	1.80	6.20
1	32924EQ111	BWRG-024-S3BA,CA,DA	20,400	7.70	20,400	2.50	12,000	1.80	5.80
1	32924EQ111	BWRE-024-S7BA,DA	21,000	7.80	23,000	2.50	13,000	1.70	5.60

AIRTEMP CORPORATION (CONT.)

Type: HRCU-A-CB Trade Name: Airtemp

	Model		Cooling		Heating				
footnote(s)	outdoor unit	indoor unit	capacity (Btuh)	SEER	capacity (Btuh)	COP @ 47°	capacity (Btuh)	COP @ 17°	HSPF
1	32924EQ111	* BWRJ-024-C2AA	22,000	8.00	23,400	2.60	13,000	1.70	5.90
1	32924EQ111	BWRE-030-S7BA,DA,EA	22,000	8.00	23,400	2.60	13,000	1.80	5.70
	32930HK111	* BWRE-030-S7BA,DA,EA	29,400	7.80	30,400	2.60	17,500	1.80	5.90
	32930HK111								
	32930HK111								
		BWRJ-036-C2AA	30,400	8.00	32,000	2.60	18,000	1.90	6.10
		BWRE-036-S7DA,EA,FA	31,000	8.00	31,000	2.70	18,500	1.90	6.20
1	32936HQ111	* BWRJ-036-C2AA	34,000	8.00	35,000	2.50	21,400	1.90	6.00
1	32936HQ111	BWRE-036-S7DA,EA,FA	34,400	8.00	36,000	2.60	22,000	2.00	6.30
1	32942EQ111	* BWRJ-048-C2AA	39,000	8.00	40,000	2.60	24,400	1.90	6.10
1	32942EQ111	BWRE-048-S7EA,GA,HA	39,000	8.00	40,000	2.60	24,400	2.00	6.30
1	32948EQ111	* BWRJ-048-C2AA	44,000	8.00	51,000	2.50	27,000	1.60	5.40
1	32948EQ111	BWRE-048-S7EA,GA,HA	44,000	8.00	46,000	2.50	27,000	1.80	5.80
	32960EK111	* BWRJ-060-C2AA	54,000	8.00	55,000	2.70	37,000	2.10	6.50
	32960EK111	BWRE-060-S7EA,GA,HA	54,000	8.00	55,000	2.70	37,000	2.10	6.50

AIRTEMP CORPORATION (CONT.)

Type: HRCU-A-C Trade Name-All Seasons

	Model		Cooling		Heating				
footnote(s)	outdoor unit	indoor unit	capacity (Btuh)	SEER	capacity (Btuh)	COP @ 47°	capacity (Btuh)	COP @ 17°	HSPF
1	32918EQ111	COLH-018-SCOA	17,000	8.20	17,000	2.80	10,500	1.60	6.10
1	32924EQ111	COLH-024-ACOA	20,400	8.00	22,400	2.50	13,000	1.80	5.80
	32930HK111	COLH-036-AGOA	30,400	8.30	30,400	2.50	18,500	1.90	6.00
1	32936HQ111	COLH-036-ACOA	35,400	8.00	36,400	2.70	21,400	1.90	6.20
1	32936HQ111	COLH-038-ACOA	35,400	8.00	36,400	2.70	21,400	1.90	6.20
1	32942EQ111	COLH-048-ACOA	38,000	8.00	40,000	2.50	24,000	1,90	6.00
1	32948EQ111	COLH-048-ACOA	42,000	7.50	45,000	2.40	26,000	1.70	5.40
1	32960EK111	COLH-060-ACOA	52,000	7.50	53,000	2.50	36,000	2.00	6.10

Products not Covered by DOE

Type: HSP-A Trade Name: Airtemp

footnote(s)	outdoor unit	indoor unit	capacity (Btuh)	EER	capacity (Btuh)	COP @ 47°	capacity (Btuh)	COP @ 17°	HSPF
1	31030CR111	...	35,400	7.50	38,000	2.70	18,500	1.60	...
1	31040DR111	...	46,000	7.50	48,000	2.50	27,400	1.60	...
1	31050DR111	...	57,000	7.50	60,000	2.50	35,000	1.70	...
1	31075ER212	...	86,000	7.50	88,000	2.60	48,000	1.50	...
1	31075EP212	...	86,000	7.50	88,000	2.60	48,000	1.50	...

AIRTEMP CORPORATION (CONT.)

Type: HSP-A Trade Name: Airtemp

	Model		Cooling		Heating				
footnote(s)	outdoor unit	indoor unit	capacity (Btuh)	EER	capacity (Btuh)	COP @ 47°	capacity (Btuh)	COP @ 17°	HSPF
1	31075ER211	...	86,000	7.50	88,000	2.60	48,000	1.50	...
1	31075EP211	...	86,000	7.50	88,000	2.60	48,000	1.50	...
1	31100EP212	...	113,000	7.50	114,000	2.60	63,000	1.70	...
1	31100ER212	...	113,000	7.50	114,000	2.60	63,000	1.70	...
1	31100ER211	...	113,000	7.50	114,000	2.60	63,000	1.70	...
1	31100EP211	...	113,000	7.50	114,000	2.60	63,000	1.70	...

Type: HRCU-A-CB Trade Name: Airtemp

	Model		Cooling		Heating				
footnote(s)	outdoor unit	indoor unit	capacity (Btuh)	EER	capacity (Btuh)	COP @ 47°	capacity (Btuh)	COP @ 17°	HSPF
1	32948ER111	BWRJ-048-C2AA	44,000	7.50	51,000	2.50	27,000	1.60	...
1	32948ER111	BWRE-048-S7EA,GA,HA	44,000	7.50	46,000	2.50	27,000	1.80	...
1	32960ER111	BWRJ-060-C2AA	54,000	7.90	55,000	2.70	37,000	2.10	...
1	32960ER111	BWRE-060-S7EA,GA,HA	54,000	7.90	55,000	2.70	37,000	2.10	...

AIRTEMP CORPORATION (CONT.)

Type: HRCU-A-C Trade Name-All Seasons

	Model		Cooling		Heating				
footnote(s)	outdoor unit	indoor unit	capacity (Btuh)	EER	capacity (Btuh)	COP @ 47°	capacity (Btuh)	COP @ 17°	HSPF
1	32948ER111	COLH-048-ACOA	42,000	7.10	45,000	2.40	26,000	1.70	...
1	32960ER111	COLH-060-ACOA	52,000	7.00	53,000	2.50	36,000	2.00	...

CARRIER AIR CONDITIONING
DOE Covered Products

Type: HSP-A Trade Name: Single Package Heat Pump

	Model		Cooling			Heating			
footnote(s)	outdoor unit	indoor unit	capacity (Btuh)	SEER	capacity (Btuh)	COP @ 47°	capacity (Btuh)	COP @ 17°	HSPF
.	50YQ02431	...	24600	7.60	26,200	2.65	13,500	1.70	5.85
.	50YQ03031	...	28,600	8.10	30,200	2.80	15,000	1.70	5.95
.	50YQ03631	...	33,800	8.00	35,400	2.60	18,600	1.70	5.90
.	50YQ04231	...	40,000	8.00	41,000	2.55	22,200	1.65	5.70
.	50YQ04831	...	47,000	8.10	50,500	2.65	27,800	1.80	6.10
.	50YQ06031	...	54,500	7.50	59,500	2.60	32,200	1.80	5.95

Type: HRCU-A-C Trade Name: Compact Heat Pump

footnote(s)	outdoor unit	indoor unit	capacity (Btuh)	SEER	capacity (Btuh)	COP @ 47°	capacity (Btuh)	COP @ 17°	HSPF
.	38CQ01534	28VQ018	14,900	7.10	16,000	2.35	8,700§	1.40§	5.40
.	WAS	...	14,900	7.10	16,000	2.35	9,100	1.50	5.40
.	38CQ02034	28HQ,VQ024	16,300§	6.50§	16,800§	2.20§	9,700§	1.55§	5.25§
.	WAS	...	17,800	7.50	18,500	2.45	10,800	1.75	5.60
.	38CQ02735	28HQ/VQ024	24,000	6.55	26,000	2.40	14,900	1.70	6.05
.	38CQ02735	28HQ/VQ030	24,800	6.75	27,000	2.50	15,400	1.75	6.00
.	38CQ03334	28HQ,VQ030	29,000	6.70	32,200	2.40	19,400	1.80	5.75

CARRIER AIR CONDITIONING (CONT.)

Type: HRCU-A-C Trade Name: Compact Heat Pump

footnote(s)	Model		Cooling		Heating				
	outdoor unit	indoor unit	capacity (Btuh)	SEER	capacity (Btuh)	COP @ 47°	capacity (Btuh)	COP @ 17°	HSPF
.	38CQ03334	28HQ,VQ036	31,200	7.00	34,600	2.55	20,600	1.85	5.75
.	38CQ03934	28HQ,VQ036	34,600	6.85	37,600	2.50	23,800	1.90	6.10
.	38CQ03934	28HQ,VQ042	34,800	6.90	37,800	2.50	24,000	1.95	6.10
.	38CQ04434	28HQ,VQ048	42,500	7.15	47,500	2.45	29,200	1.95	6.55

Type: HRCU-A-CB Trade Name: Compact Heat Pump

footnote(s)	outdoor unit	indoor unit	capacity (Btuh)	SEER	capacity (Btuh)	COP @ 47°	capacity (Btuh)	COP @ 17°	HSPF
.	38CQ01534	40DQ014	13,700	6.70	15,000	2.20	8,200§	1.35§	5.40
.	WAS	...	13,700	6.70	15,000	2.20	8,600	1.45	5.40
.	38CQ01534	* 40AQ018	14,700	6.65	16,200	2.30	8,800§	1.45§	5.40
.	WAS	...	14,700	6.65	16,200	2.30	9,300	1.55	5.40
.	38CQ0.1534	40DQ018	15,000	6.95	15,600	2.25	8,900§	1.50§	5.40
.	WAS	...	15,000	6.95	15,600	2.25	9,400	1.60	5.40
.	38CQ02034	* 40AQ024	16,100§	6.30§	16,900§	2.20§	9,700§	1.50§	5.25§
.	WAS	...	17,600	7.30	18,700	2.45	10,900	1.70	5.60

CARRIER AIR CONDITIONING (CONT.)

Type: HRCU-A-C Trade Name: Compact Heat Pump

	Model		Cooling				Heating		
footnote(s)	outdoor unit	indoor unit	capacity (Btuh)	SEER	capacity (Btuh)	COP @ 47°	capacity (Btuh)	COP @ 17°	HSPF
.	38CQ02034	40DQ018	16,200§	6.45§	16,700§	2.20§	9.600§	1.50§	5.25§
.	WAS	...	17,700	7.45	18,400	2.45	10,700	1.70	5.60
.	38CQ02034	40DQ024	17,000§	6.80§	17,500§	2.35§	10,000§	1.65§	5.25§
.	WAS	...	18,600	7.85	19,300	2.60	11,200	1.85	5.60
.	38CQ02735	40AQ024	23,800	6.45	25,800	2.35	14,800	1.70	5.80
.	38CQ02735	40DQ024	24,200	6.70	26,000	2.40	14,800	1.75	5.95
.	38CQ02735	* 40AQ030	24,800	6.80	27,000	2.45	15,400	1.70	6.00
.	38CQ02735	40DQ030	25,600	7.05	27,600	2.55	15,800	1.85	6.25
.	38CQ03334	40AQ030	29,000	6.60	32,200	2.45	19,400	1.80	5.75
.	38CQ03334	* 40AQ036	31,000	6.90	34,400	2.50	20,600	1.85	5.75
4	38CQ03334	28HQ,VQ036W/40FS160	31,000	6.90	34,400	2.50	20,600	1.80	5.75
4	38CQ03934	28HQ,VQ036W/40FS160	34,400	6.85	37,600	2.50	23,800	1.90	6.10
.	38CQ03934	40AQ036	34,400	6.85	37,600	2.50	23,800	1.90	6.10
4	38CQ03934	* 28HQ,VQ042W/40FS160	34,800	6.90	37,800	2.50	24,000	1.90	6.10
.	38CQ03934	40QB042	36,200	7.15	37,000	2.45	23,400	1.85	6.20

CARRIER AIR CONDITIONING (CONT.)

Type: HRCU-A-CB Trade Name: Compact Heat Pump

	Model		Cooling			Heating			
footnote(s)	outdoor unit	indoor unit	capacity (Btuh)	SEER	capacity (Btuh)	COP @ 47°	capacity (Btuh)	COP @ 17°	HSPF
.	38CQ04434	* 28HQ,VQ048W/40FS200	42,500	7.15	47,500	2.45	29,200	1.95	6.55
.	38CQ04434	40QB042	43,500	7.05	48,500	2.55	29,000	1.85	6.15
.	38CQ04434	40QB048	43,500	7.05	48,500	2.55	29,000	1.85	6.15

Type: HRCU-A-C Trade Name: Weathermaster III

footnote(s)	outdoor unit	indoor unit	capacity (Btuh)	SEER	capacity (Btuh)	COP @ 47°	capacity (Btuh)	COP @ 17°	HSPF
4	38HQ940	38HQ12031+28HQ,VQ030	21,200	7.85	22,200	2.55	12,000	1.65	5.65
4	39HQ940	38HQ12031+29HQ,VQ036	21,400	8.10	22,400	2.60	12,000	1.65	5.65
4	38HQ940	38HQ12731+28HQ,VQ030	27,400	8.00	29,200	2.60	15,700§	1.95§	6.50
.	WAS	...	27,400	8.00	29,200	2.60	16,800	1.85	6.50
4	38HQ940	38HQ12731+28HQ,VQ036	27,800	8.25	29,600	2.50	16,400§	1.85§	6.10
.	WAS	...	27,800	8.25	29,600	2.50	17,500	1.75	6.10
4	38HQ940	38HQ13431+28HQ,VQ036	33,800	7.85	37,600	2.70	21,400§	1.95§	6.45
.	WAS	...	33,800	7.85	37,600	2.70	23,000	2.10	6.45

CARRIER AIR CONDITIONING (CONT.)

Type: HRCU-A-C Trade Name: Weathermaster III

	Model		Cooling			Heating			
footnote(s)	outdoor unit	indoor unit	capacity (Btuh)	SEER	capacity (Btuh)	COP @ 47°	capacity (Btuh)	COP @ 17°	HSPF
4	38HQ960	38HQ14031+28HQ,VQ042	41,500	8.00	44,500	2.65	26,200	1.95	6.60
4	38HQ960	38HQ14631+28HQ,VQ048	44,000	8.00	48,500	2.65	29,800	2.00	6.70

Type: HRCU-A-CB Trade Name: Weathermaster III

footnote(s)	outdoor unit	indoor unit	capacity (Btuh)	SEER	capacity (Btuh)	COP @ 47°	capacity (Btuh)	COP @ 17°	HSPF
4	38HQ940W/38HQ12031	28HQ,VQ036W/40FS160	20,800	8.05	22,400	2.50	12,000	1.60	5.65
4	38HQ940	* 38HQ12031+40AQ030	21,000	7.70	22,200	2.50	12,000	1.65	5.65
4	38HQ940	38HQ12031+40AQ036	21,200	8.00	22,400	2.50	12,000	1.60	5.65
4	38HQ940	* 38HQ12731+40AQ030	27,200	8.00	29,400	2.55	15,900§	1.90§	6.45
•	WAS	...	27,200	8.00	29,400	2.55	17,000	1.80	6.45
4	38HQ940	38HQ12731+49DQ030	27,400	8.00	30,200	2.65	16,700§	2.05§	6.50
•	WAS	...	27,400	8.00	30,200	2.65	17,900	1.95	6.50
4	38HQ940W/38HQ12731	28HQ,VQ036W/40FS160	27,600	8.05	29,600	2.50	16,400§	1.85§	6.10
•	WAS	...	27,600	8.05	29,600	2.50	17,500	1.75	6.10
4	38HQ940	38HQ12731+40AQ036	27,600	8.05	29,600	2.50	16,400§	1.85§	6.10

CARRIER AIR CONDITIONING (CONT.)

Type: HRCU-A-CB Trade Name: Weathermaster III

	Model						Heating		
footnote(s)	outdoor unit	indoor unit	capacity (Btuh)	SEER	capacity (Btuh)	COP @ 47°	capacity (Btuh)	COP @ 17°	HSPF
.	WAS	...	27,600	8.05	29,600	2.50	17,500	1.75	6.10
4	38HQ940W/38HQ13431	28HQ,VQ036W/40FS160	33,600	7.80	37,800	2.65	21,600§	1.90§	6.45
.	WAS	...	33,600	7.80	37,800	2.65	23,200	2.05	6.45
4	38HQ940	* 38HQ13431+40AQ036	33,600	7.80	37,800	2.65	21,600§	1.90§	6.45
.	WAS	...	33,600	7.80	37,800	2.65	23,200	2.05	6.45
4	38HQ960	38HQ14031+40QB042	39,500	8.00	44,000	2.55	25,200	1.90	6.30
4	38HQ960W/38HQ14031	* 28HQ,VQ042W/40FS160	41,000	8.05	44,500	2.60	26,600	1.95	6.70
4	38HQ960W/38HQ14031	28HQ,VQ042W/40FS200	41,000	8.00	45,000	2.60	26,600	1.90	6.60
4	38HQ960W/38HQ14631	28HQ,VQ048W/40FS220	43,500	7.70	48,500	2.60	29,800	2.30	6.70
4	38HQ960W/38HW14631	* 28HQ,VQ048W/40FS200	44,000	8.00	48,000	2.65	29,400	2.00	6.70
4	38HQ960	38HQ14631+40QB048	45,000	8.00	48,500	2.60	29,600	1.95	6.50

Type: HRCU-A-C Trade Name: Weathermaster Plus

footnote(s)	outdoor unit	indoor unit	capacity (Btuh)	SEER	capacity (Btuh)	COP @ 47°	capacity (Btuh)	COP @ 17°	HSPF
4	38HQ940	38QW02730+28HQ/VQ030	27,400	8.00	29,200	2.60	16,800	1.85	6.50

CARRIER AIR CONDITIONING (CONT.)

Type: HRCU-A-C Trade Name: Weathermaster Plus

	Model		Cooling		Heating				
footnote(s)	outdoor unit	indoor unit	capacity (Btuh)	SEER	capacity (Btuh)	COP @ 47°	capacity (Btuh)	COP @ 17°	HSPF
4	38HQ940	38QW02730+28HQ/VQ036	27.800	8.25	29,600	2.50	16,300	1.85	6.10
4	38HQ940	38QW03430+28HQ/VQ036	33,800	7.85	37,600	2.70	21,400	1.95	6.45
4	38HQ960	38QW04030+28HQ/VQ042	41,500	8.00	44,500	2.65	26,200	1.95	6.60
4	38HQ960	38QW04630+28HQ/VQ048	44,000	8.00	48,500	2.65	29,800	2.00	6.70

Type: HRCU-A-CB Trade Name: Weathermaster Plus

footnote(s)	outdoor unit	indoor unit	capacity (Btuh)	SEER	capacity (Btuh)	COP @ 47°	capacity (Btuh)	COP @ 17°	HSPF
4	38HQ940	* 38QW02730+40AQ030	27,200	8.00	29,400	2.55	17,000	1.80	6.45
4	38HQ940	38QW02730+40DQ030	27,400	8.00	30,200	2.65	17,900	1.95	6.50
4	38HQ940	38QW02730+40AQ036	27,600	8.05	29,600	2.50	17,500	1.75	6.10
4	38HQ940/38QW02730	28HQ,VQ036W/40FS160	27,600	8.05	29,600	2.50	17,500	1.75	6.10
4	38HQ940	* 38QW03430+40AQ036	33,600	7.80	37,800	2.65	23,200	2.05	6.45
4	38HQ940/38QW03430	28HQ,VQ036W/40FS160	33,600	7.80	37,800	2.65	23,200	2.05	6.45
4	38HQ960	38QW03040+40QB042	39,500	8.00	44,000	2.55	25,200	1.90	6.30
4	38HQ960/38QW04030	* 28HQ,VQ042W/40FS160	41,000	8.05	44,500	2.60	26,600	1.95	6.70
4	38HQ960/38QW04030	28HQ,VQ042W/40FS200	41,000	8.05	45,000	2.60	26,600	1.90	6.60

CARRIER AIR CONDITIONING (CONT.)

Type: HRCU-A-CB Trade Name: Weathermaster Plus

	Model		Cooling			Heating			
footnote(s)	outdoor unit	indoor unit	capacity (Btuh)	SEER	capacity (Btuh)	COP @ 47°	capacity (Btuh)	COP @ 17°	HSPF
4	38HQ960/38QW04630	28HQ,VQ048W/40FS220	43,500	7.70	48,500	2.60	29,800	2.30	6.70
4	38HQ960/38QW04630	* 28HQ,VQ048W/40FS200	44,000	8.00	48,000	2.65	29,400	2.00	6.70
4	38HQ960	38QW04630+40QB048	45,000	8.00	48,500	2.60	29,600	1.95	6.50

Type: HRCU-A-C Trade Name: Year Round One

footnote(s)	outdoor unit	indoor unit	capacity (Btuh)	SEER	capacity (Btuh)	COP @ 47°	capacity (Btuh)	COP @ 17°	HSPF
.	38QB01530	28VQ018	13,900	8.00	14,100	2.50	7,500	1.50	5.65
.	38QB01831	28HQ/VQ024	18,200	8.20	18,900	2.45	10,500	1.60	5.70
.	38QB02430	28HQ/VQ024	23,200	7.60	24,600	2.40	13,200	1.55	5.65
.	38QB02430	28HQ/VQ030	24,600	8.05	25,600	2.50	14,200	1.65	5.90
.	38QB03030	28HQ/VQ030	28,600	7.70	28,400	2.50	15,400	1.65	5.60
.	38QB03030	28HQ/VQ036	29,600	8.10	29,000	2.55	16,000	1.70	5.75

CARRIER AIR CONDITIONING (CONT.)

Type: HRCU-A-C Trade Name: Year Round One

	Model		Cooling		Heating				
footnote(s)	outdoor unit	indoor unit	capacity (Btuh)	SEER	capacity (Btuh)	COP @ 47°	capacity (Btuh)	COP @ 17°	HSPF
.	38QB03630	28HQ/VQ036	35,000	8.10 .	34,800	2.60	20,600	1.95	6.35
.	38QB03630	28HQ/VQ042	35,600	8.15	35,600	2.55	21,400	1.95	6.35
.	38QB04230	28HQ/VQ042	41,000	8.05	40,500	2.50	23,800	1.85	6.15
.	38QB04830	28HQ/VQ048	47,000	8.15	45,500	2.60	26,800	1.90	6.35

Type: HRCU-A-CB Trade Name: Year Round One

footnote(s)	outdoor unit	indoor unit	capacity (Btuh)	SEER	capacity (Btuh)	COP @ 47°	capacity (Btuh)	COP @ 17°	HSPF
.	38QB01530	40DQ018	13,600	7.70	13,900	2.40	7,500	1.45	5.40
.	38QB01530	* 40AQ018	13,900	8.00	14,100	2.50	7,500	1.50	5.65
.	38QB01831	* 40AQ024	18,100	8.10	19,000	2.50	10,600	1.60	5.70
.	38QB01831	40DQ024	18,300	8.50	18,600	2.60	10,400	1.65	5.75
.	38QB02430	40AQ024	22, 800	7.25	24,800	2.35	13,600	1.55	5.60
.	38QB02430	40DQ024	23,000	7.55	24,400	2.40	13,200	1.55	5.65
.	38QB02430	* 40AQ030	24,600	8.05	25,600	2.50	14,200	1.65	5.90
.	38QB02430	40DQ030	25,000	8.00	26,200	2.60	14,500	1.70	6.00

CARRIER AIR CONDITIONING (CONT.)

Type: HRCU-A-CB Trade Name: Year Round One

	Model		Cooling		Heating				
footnote(s)	outdoor unit	indoor unit	capacity (Btuh)	SEER	capacity (Btuh)	COP @ 47°	capacity (Btuh)	COP @ 17°	HSPF
.	38QB03030	40AQ030	28,400	7.55	28,600	2.45	15,500	1.65	5.55
.	38QB03030	40DQ030	29,000	7.55	29,000	2.50	15,800	1.65	5.60
.	38QB03030	* 40AQ036	29,400	8.05	29,200	2.60	16,000	1.70	5.75
4	38QB03030	28HQ,VQ036W/40FS160	29,400	8.05	29,200	2.55	16,000	1.70	5.75
.	38QB03030	40AQ036	34,800	8.00	35,000	2.55	20,800	1.90	6.25
4	38QB03630	28HQ,VQ036W/40FS160	34,800	8.00	35,000	2.55	20,800	1.90	6.25
.	38QB03630	28HQ,VQ042W/40FS160	35,400	8.05	35,800	2.55	21,600	1.95	6.35
.	38QB03630	* 40QB042	36,200	8.25	36,600	2.65	21,600	2.00	6.50
4	38QB04230	28HQ,VQ042W/40FS160	41,000	7.90	40,500	2.50	23,800	1.85	6.15
.	38QB04230	* 40QB042	41,500	8.10	41,500	2.60	24,200	1.90	6.35
4	38QB04830	28HQ,VQ048W/40FS200	47,000	8.05	45,500	2.55	27,000	1.90	6.35
.	38QB04830	* 40QB048	47,000	8.05	47,500	2.70	28,400	2.00	6.60
.	38QB06030	* 40QB060	53,000	7.50	53,000	2.50	31,400	1.85	6.20

CARRIER AIR CONDITIONING (CONT.)
Products not Covered by DOE

Type: HSP-A Trade Name: Single Package Heat Pump

	Model		Cooling		Heating				
footnote(s)	outdoor unit	indoor unit	capacity (Btuh)	EER	capacity (Btuh)	COP @ 47°	capacity (Btuh)	COP @ 17°	HSPF
.	50YQ0365	...	32,800	7.35	35,400	2.60	17,100	1.55	...
.	50YQ0424,5,6	...	40,000	7.25	41,000	2.55	20,400	1.50	...
.	50YQ0484,5,6	...	47,000	7.20	50,500	2.65	25,600	1.65	...

Type: HRCU-A-C Trade Name: Compact Heat Pump

footnote(s)	outdoor unit	indoor unit	capacity (Btuh)	EER	capacity (Btuh)	COP @ 47°	capacity (Btuh)	COP @ 17°	HSPF
.	38CQ0335	28HQ,VQ030	29,000	6.25	32,200	2.40	17,800	1.65	...
.	38CQ0335	28HQ,VQ036	31,200	6.60	34,600	2.50	19,000	1.70	...
.	38CQ0394,5	28HQ,VQ036	34,600	6.35	37,600	2.50	22,000	1.75	...

CARRIER AIR CONDITIONING (CONT.)

Type: HRCU-A-C Trade Name: Compact Heat Pump

	Model		Cooling			Heating			
footnote(s)	outdoor unit	indoor unit	capacity (Btuh)	EER	capacity (Btuh)	COP @ 47°	capacity (Btuh)	COP @ 17°	HSPF
.	38CQ0394,5	28HQ,VQ042	34,800	6.50	37,800	2.50	22,000	1.75	...
.	38CQ0444,5	28HQ,VQ048	42,500	6.50	47,500	2.45	26,800	1.75	...

Type: HRCU-A-CB Trade Name: Compact Heat Pump

footnote(s)	outdoor unit	indoor unit	capacity (Btuh)	EER	capacity (Btuh)	COP @ 47°	capacity (Btuh)	COP @ 17°	HSPF
.	38CQ0335	40AQ030	29,000	6.25	32,400	2.45	17,800	1.65	...
4	38CQ0335	28HQ,VQ036W/40FS160	31,000	6.45	34,400	2.50	19,000	1.70	...
.	38CQ0335	40AQ036	31,000	6.45	34,400	2.50	19,000	1.70	...
4	38CQ0394,5	28HQ,VQ036W/49FS160	34,400	6.30	37,600	2.50	21,800	1.75	...
.	38CQ0394,5	40AQ036	34,400	6.30	37,600	2.50	21,800	1.75	...
4	38CQ0394,5	28HQ,VQ042W/40FS160	34,800	6.45	37,800	2.50	22,000	1.75	...
.	38CQ0394,5	40QB042	36,200	6.65	37,000	2.45	21,600	1.70	...
4	38CQ0444,5	28HQ,VQ048W/40FS200	42,500	6.50	47,500	2.45	26,800	1.70	...
.	38CQ0444,5	40QB042	43,500	6.55	48,500	2.55	26,800	1.80	...
.	38CQ0444,5	40QB048	43,500	6.55	48,500	2.55	26,600	1.70	...

CARRIER AIR CONDITIONING (CONT.)

Type: HSP-A Trade Name: Commercial Heat Pump Weathermaker

	Model		Cooling			Heating			
footnote(s)	outdoor unit	indoor unit	capacity (Btuh)	EER	capacity (Btuh)	COP @ 47°	capacity (Btuh)	COP @ 17°	HSPF
2	50QD0063	...	59,000	7.50	60,000	2.70	36,000	2.10	...
.	50QH0063	...	59,000	7.50	60,000	2.70	36,000	2.10	...
2	50QD0065,6	...	60,000	7.60	60,000	2.70	36.000	2.10	...
.	50QH0065,6	...	60,000	7.60	60,000	2.70	36,000	2.10	...
.	50PQ0081,4,5,6	...	87,000	7.60	92,000	2.80	52,000	2.00	...
.	50RQ0081,4,5,6	...	88,000	7.80	91,000	2.80	51,000	2.00	...
.	50PQ0101,4,5,6	...	107,000	7.50	113,000	2.75	62,000	2.00	...
.	50RQ0101,5,6	...	108,000	7.70	112,000	2.80	61,000	2.00	...
.	50PQ0121,4,5,6	...	120,000	8.20	120,000	2.85	65,000	2.10	...

Type: HRCU-A-CB Trade Name: Commercial Heat Pump Weathermaker

footnote(s)	outdoor unit	indoor unit	capacity (Btuh)	EER	capacity (Btuh)	COP @ 47°	capacity (Btuh)	COP @ 17°	HSPF
.	38DQ0085,6	40BA009	86,000	9.05	86,000	3.15	47,000	2.10	...
.	38AQ0125,6	40RR012	117,000	9.10	117,000	3.00	69,000	2.20	...

CARRIER AIR CONDITIONING (CONT.)

Type: HRCU-A-C Trade Name: Year Round One

	Model		Cooling			Heating			
footnote(s)	outdoor unit	indoor unit	capacity (Btuh)	EER	capacity (Btuh)	COP @ 47°	capacity (Btuh)	COP @ 17°	HSPF
.	38QB0365,6	28HQ/VQ036	35,000	7.95	34,800	2.60	20,600	1.95	...
.	38QB0365,6	28HQ/VQ042	35,600	7.90	35,600	2.55	21,400	1.95	...
.	38QB0425,6	28HQ/VQ042	41,000	7.55	40,500	2.50	23,800	1.85	...
4	38QB0425,6	28HQ,VQ042W	41,000	7.50	40,500	2.50	23,800	1.85	...
.	38QB0485,6	28HQ/VQ048	47,000	7.75	45,500	2.60	26,800	1.90	...

Type: HRCU-A-CB Trade Name: Year Round One

footnote(s)	outdoor unit	indoor unit	capacity (Btuh)	EER	capacity (Btuh)	COP @ 47°	capacity (Btuh)	COP @ 17°	HSPF
.	38QB0365,6	40AQ036	34,800	7.70	35,000	2.55	20,800	1.90	...
4	38QB0365,6	28HQ,VQ036W/40FS160	34,800	7.70	35,000	2.55	20,800	1.90	...
4	38QB0365,6	28HQ,VQ042W/40FS160	35,400	7.85	35,800	2.55	21,600	1.95	...
.	38QB0365,6	40QB042	36,200	7.95	36,600	2.65	21,600	2.00	...
.	38QB0425,6	40Q B042	41,500	7.65	41,500	2.60	24,200	1.90	...
4	38QB0485,6	28HQ,VQ048W/40FS200	47,000	7.70	45,500	2.55	27,000	1.90	...

CARRIER AIR CONDITIONING (CONT.)

Type: HRCU-A-CB Trade Name: Year Round One

	Model		Cooling		Heating				
footnote(s)	outdoor unit	indoor unit	capacity (Btuh)	SEER	capacity (Btuh)	COP @ 47°	capacity (Btuh)	COP @ 17°	HSPF
·	38QB0485,6	40QB048	47,000	7.65	47,500	2.70	28,400	2.00	...
·	38QB0605,6	40QB060	53,000	7.10	53,000	2.50	31,400	1.85	...

COLEMAN COMPANY, INC.
DOE Covered Products

Type: HSP-A Trade Name: T.H.E. Heat Pump

Model			Cooling			Heating			
footnote(s)	outdoor unit	indoor unit	capacity (Btuh)	SEER	capacity (Btuh)	COP @ 47°	capacity (Btuh)	COP @ 17°	HSPF
.	6524-901	...	25,200	10.20	24,800	3.00	13,200	1.90	7.00
.	6530-901	...	30,200	9.30	30,400	2.90	16,800	2.00	6.80
.	6536-901	...	35,600	9.45	37,200	3.05	19,600	2.00	7.10
.	6542-901	...	40,000	9.20	43,500	3.10	22,000	2.00	7.15

Type: HRCU-A-C Trade Name: T.H.E. Heat Pump

footnote(s)	outdoor unit	indoor unit	capacity (Btuh)	SEER	capacity (Btuh)	COP @ 47°	capacity (Btuh)	COP @ 17°	HSPF
.	3318B601	*3318A830	19,200	11.00	19,600	3.25	11,600§	2.30	8.00
.	3318B611	3318-833	19,200	11.00	19,600	3.25	11,600§	2.30	8.00
.	3224-601	*3224-830	24,600§	8.50	28,600§	2.70	15,500	2.00	6.80
.	3224A901	*3324-831	25,400	9.00	27,600	2.90	15,900	2.10	7.00
.	3324B901	3324-830	25,400	9.00	27,600	2.90	15,900	2.10	7.00
.	3324B901	3324A830	25,400	9.00	27,600	2.90	15,900	2.10	7.00
.	3324B901	3324-832	25,400	9.00	27,600	2.90	15,900	2.10	7.00
.	3324B911	3324-833	25,400	9.00	27,600	2.90	15,900	2.10	7.00
.	3324B911	3324-823	25,400	9.00	27,600	2.90	15,900	2.10	7.00

COLEMAN COMPANY, INC. (CONT.)

Type: HRCU-A-C Trade Name: T.H.E. Heat Pump

Model			Cooling		Heating				
footnote(s)	outdoor unit	indoor unit	capacity (Btuh)	SEER	capacity (Btuh)	COP @ 47°	capacity (Btuh)	COP @ 17°	HSPF
.	3324A901	*3324A830	25,400	9.00	-27,600	2.90	15,900	2.10	7.00
.	3324A901	3324-832	25,400	9.00	27,600	2.90	15,900	2.10	7.00
.	3230D901	*3240-830	33,200	9.00	35,200	3.00	21,600	2.25	7.35
.	3230D901	3240-831,32	33,200	9.00	35,200	3.00	21,600	2.25	7.35
.	3330A901	*3340-830	33,200	9.00	35,200	3.00	21,600	2.25	7.35
.	3330A901	3340-832	33,200	9.00	35,200	3.00	21,600	2.25	7.35
.	3330B901	3340-830	33,200	9.00	35,200	3.00	21,600	2.25	7.35
.	3330B901	3340A830	33,200	9.00	35,200	3.00	21,600	2.25	7.35
.	3330B901	3340-832	33,200	9.00	35,200	3.00	21,600	2.25	7.35
.	3330B911	3340-823,33	33,200	9.00	35,200	3.00	21,600	2.25	7.35
.	3230D901	3340-830	33,200	9.00	35,200	3.00	21,600	2.25	7.35
.	3230D901	3340-832	33,200	9.00	35,200	3.00	21,600	2.25	7.35
.	3230D901	3440-830	33,200	9.00	35,200	3.00	21,600	2.25	7.35
.	3236D901	*3240-830	35,400	8.15	37,600	2.80	22,600	2.10	7.00
.	3236D901	3240-831,32	35,400	8.15	37,600	2.80	22,600	2.10	7.00

COLEMAN COMPANY, INC. (CONT.)

Type: HRCU-A-C Trade Name: T.H.E. Heat Pump

	Model		Cooling		Heating				
footnote(s)	outdoor unit	indoor unit	capacity (Btuh)	SEER	capacity (Btuh)	COP @ 47°	capacity (Btuh)	COP @ 17°	HSPF
.	3336A901	* 3340-830	35,400	8.15	37,600	2.80	22,600	2.10	7.00
.	3336A901	3340-832	35,400	8.15	37,600	2.80	22,600	2.10	7.00
.	3336A901	3340A830	35,400	8.15	37,600	2.80	22,600	2.10	7.00
.	3336B901	3340A830	35,400	8.15	37,600	2.80	22,600	2.10	7.00
.	3336B901	3340-830,32	35,400	8.15	37,600	2.80	22,600	2.10	7.00
.	3336B911	3340-823,33	35,400	8.15	37,600	2.80	22,600	2.10	7.00
.	3336D901	3340-830	35,400	8.15	37,600	2.80	22,600	2.10	7.00
.	3236D901	3340-832	35,400	8.15	37,600	2.80	22,600	2.10	7.00
.	3236D901	3340-830	35,400	8.15	37,600	2.80	22,600	2.10	7.00
.	3237D601	* 3240-830	37,200	8.65	40,000	2.90	23,000	2.20	7.05
.	3237D601	3240-831,32	37,200	8.65	40,000	2.90	23,000	2.20	7.05
.	3337B601	3340-830,32	37,200	8.65	40,000	2.90	23,000	2.20	7.05
.	3337B611	3340-823,33	37,200	8.65	40,000	2.90	23,000	2.20	7.05
.	3337A601	* 3340-830	37,200	8.65	40,000	2.90	23,000	2.20	7.05
.	3337A601	3340-832	37,200	8.65	40,000	2.90	23,000	2.20	7.05

COLEMAN COMPANY, INC. (CONT.)

Type: HRCU-A-C Trade Name: T.H.E. Heat Pump

	Model		Cooling		Heating				
footnote(s)	outdoor unit	indoor unit	capacity (Btuh)	SEER	capacity (Btuh)	COP @ 47°	capacity (Btuh)	COP @ 17°	HSPF
.	3337A601	3340A830	37,200	8.65	40,000	2.90	23,000	2.20	7.05
.	3337B601	3340A830	37,200	8.65	40,000	2.90	23,000	2.20	7.05
.	3237D601	3340-830	37,200	8.65	40,000	2.90	23,000	2.20	7.05
.	3237D601	3440-830	37,200	8.65	40,000	2.90	23,000	2.20	7.05
.	3237D601	3340-832	37,200	8.65	40,000	2.90	23,000	2.20	7.05
.	3240D901	*3240-830	38,500	8.60§	42,000	2.85	26,000§	2.25§	7.10
.	3340B901	3340-830,32	38,500	8.60	42,000	2.85	26,000	2.25	7.40
.	3340B911	3340-823,33	38,500	8.60	42,000	2.85	26,000	2.25	7.40
.	3340A901	*3340-830	38,500	8.60	42,000	2.85	26,000	2.25	7.40
.	3340A901	3340-832	38,500	8.60	42,000	2.85	26,000	2.25	7.40
.	3340A901	3340A830	38,500	8.60	42,000	2.85	26,000	2.25	7.40
.	3340B901	3340A830	38,500	8.60	42,000	2.85	26,000	2.25	7.40
.	3240D901	3340-830	38,500	8.60	42,000	2.85	26,000	2.25	7.40
.	3240D901	3340-832	38,500	8.60	42,000	2.85	26,000	2.25	7.40
.	3240D901	3440-830	38,500	8.60	42,000	2.85	26,000	2.25	7.40

COLEMAN COMPANY, INC.

Type: HRCU-A-C Trade Name: T.H.E. Heat Pump

	Model		Cooling			Heating			
footnote(s)	outdoor unit	indoor unit	capacity (Btuh)	SEER	capacity (Btuh)	COP @ 47°	capacity (Btuh)	COP @ 17°	HSPF
.	3240D901	3240-831,32	38,500	8.60§	42,000	2.85	26,000§	2.25§	7.10
.	3243D901	*3248-830	41,500	8.60	46,000	3.05	26,800	2.35	7.50
.	3243D901	3248-831,32	41,500	8.60	46,000	3.05	26,800	2.35	7.50
.	3343A901	3348-830,32	41,500	8.60	46,000	3.05	26,800	2.35	7.50
.	3343A911	3348-832,33	41,500	8.60	46,000	3.05	26,800	2.35	7.50
.	3343A901	3348A830	41,500	8.60	46,000	3.05	26,800	2.35	7.50
.	3243D901	3348-830	41,500	8.60	46,000	3.05	26,800	2.35	7.50
.	3243D901	3348-832	41,500	8.60	46,000	3.05	26,800	2.35	7.50
.	3243D901	3448-830	41,500	8.60	46,000	3.05	26,800	2.35	7.50
.	3248D901	*3248-830	47,000	8.45	53,500	2.90	33,000	2.30	7.30
.	3248D901	3248-831,32	47,000	8.45	53,500	2.90	33,000	2.30	7.30
.	3348A901	3348-830,32	47,000	8.45	53,500	2.90	33,000	2.30	7.30
.	3348A901	3348A830	47,000	8.45	53,500	2.90	33,000	2.30	7.30
.	3348A911	3348-823,33	47,000	8.45	53,500	2.90	33,000	2.30	7.30
.	3248D901	3348-830	47,000	8.45	53,500	2.90	33,000	2.30	7.30

COLEMAN COMPANY, INC. (CONT.)

Type: HRCU-A-C Trade Name: T.H.E. Heat Pump

Model			Cooling		Heating				
footnote(s)	outdoor unit	indoor unit	capacity (Btuh)	SEER	capacity (Btuh)	COP @ 47°	capacity (Btuh)	COP @ 17°	HSPF
.	3248D901	3348-832	47,000	8.45	53,500	2.90	33,000	2.30	7.30
.	3248D901	3448-830	47,000	8.45	53,500	2.90	33,000	2.30	7.30
.	3360-601	*3360-831	56,000	8.00	61,500	2.95	36,000	2.30	7.50
.	3360A611	3360-833	56,000	8.00	61,500	2.95	36,000	2.30	7.50
.	3360-601	3360-832	57,000	8.00	63,000	3.00	37,000	2.30	7.50
.	3360A611	*3360-823	57,000	8.00	63,000	3.00	37,000	2.30	7.50

Type: HSP-A Trade Name: Super Energy Saver

Model			Cooling		Heating				
footnote(s)	outdoor unit	indoor unit	capacity (Btuh)	SEER	capacity (Btuh)	COP @ 47°	capacity (Btuh)	COP @ 17°	HSPF
.	6024-901	...	24,800	9.60	24,600	2.85	12,900	1.80	6.80
.	6030-901	...	29,400	9.15	30,200	2.85	16,700	2.00	6.70
.	6036-901	...	35,600	9.45	37,200	3.05	19,600	2.00	7.10
	6042-901	...	40,000	8.80	43,500	3.00	22,000	1.90	6.90

COLEMAN COMPANY, INC. (CONT.)

Type: HRCU-A-C Trade Name: Super Energy Saver

	Model		Cooling		Heating				
footnote(s)	outdoor unit	indoor unit	capacity (Btuh)	SEER	capacity (Btuh)	COP @ 47°	capacity (Btuh)	COP @ 17°	HSPF
.	3424-901	3424-830	25,400	9.00	27,600	2.90	15,900	2.10	7.00
.	3424-901	3424A830	25,400	9.00	27,600	2.90	15,900	2.10	7.00
.	3424A901	3424-830	25,400	9.00	27,600	2.90	15,900	2.10	7.00
.	3424A901	3424A830	25,400	9.00	27,600	2.90	15,900	2.10	7.00
.	3430-901	3440-830	33,200	9.00	35,200	3.00	21,600	2.25	7.35
.	3430-901	3440A830	33,200	9.00	35,200	3.00	21,600	2.25	7.35
.	3430A901	3440-830	33,200	9.00	35,200	3.00	21,600	2.25	7.35
.	3430A901	3440A830	33,200	9.00	35,200	3.00	21,600	2.25	7.35
.	3436-901	3440-830	35,400	8.15	37,600	2.80	22,600	2.10	7.00
.	3436-901	3440A830	35,400	8.15	37,600	2.80	22,600	2.10	7.00
.	3436A901	3440-830	35,400	8.15	37,600	2.80	22,600	2.10	7.00
.	3436A901	3440A830	35,400	8.15	37,600	2.80	22,600	2.10	7.00
.	3437-601	3440-830	37,200	8.65	40,000	2.90	23,000	2.20	7.05
.	3437-601	3440A830	37,200	8.65	40,000	2.90	23,000	2.20	7.05
.	3437A601	3440-830	37,200	8.65	40,000	2.90	23,000	2.20	7.05

COLEMAN COMPANY, INC. (CONT.)

Type: HRCU-A-C Trade Name: Super Energy Saver

	Model		Cooling		Heating				
footnote(s)	outdoor unit	indoor unit	capacity (Btuh)	SEER	capacity (Btuh)	COP @ 47°	capacity (Btuh)	COP @ 17°	HSPF
·	3437A601	3440A830	37,200	8.65	40,000	2.90	23,000	2.20	7.05
·	3440-901	3440-830	38,500	8.60	42,000	2.85	26,000	2.25	7.40
·	3440-901	3440A830	38,500	8.60	42,000	2.85	26,000	2.25	7.40
·	3440A901	3440-830	38,500	8.60	42,000	2.85	26,000	2.25	7.40
·	3440A901	3440A830	38,500	8.60	42,000	2.85	26,000	2.25	7.40
·	3443-901	3448-830	41,500	8.60	46,000	3.05	26,800	2.35	7.50
·	3443-901	3448A830	41,500	8.60	46,000	3.05	26,800	2.35	7.50
·	3448-901	3448A830	47,000	8.45	53,500	2.90	33,000	2.30	7.30
·	3448-901	3448-830	47,000	8.45	53,500	2.90	33,000	2.30	7.30

GENERAL ELECTRIC COMPANY
DOE Covered Products

Type: HSP-A Trade Name: Weathertron

	Model		Cooling		Heating				
footnote(s)	outdoor unit	indoor unit	capacity (Btuh)	SEER	capacity (Btuh)	COP @ 47°	capacity (Btuh)	COP @ 17°	HSPF
.	BWC024E	...	24,400	7.70	25,800	2.90	13,800	1.95	6.50
.	BWC024B	...	24,600	6.80	27,200	2.30	14,200	1.50	5.45
.	BWC030B	...	30,200	6.55	32,400	2.15	16,600	1.40	5.10
.	BWC030E	...	31,000	8.20	32,800	2.60	17,000	1.75	5.95
.	BWC036B1	...	35,800	6.75	39,000	2.35	19,200	1.40	5.10
.	BWC036E1	...	37,800	8.55	41,000	2.95	21,400	1.95	6.65
.	BWC042E1	...	41,000	7.15	50,000	2.70	27,000	1.90	6.25
.	BWC042C1	...	43,000	8.00	46,500	2.65	26,400	1.80	6.25
.	BWC048C1	...	48,500	7.75	48,500	2.60	28,600	1.85	6.30
.	BWC060C1G	...	55,500	8.55	55,000	2.75	32,200	1.95	6.55
.	BWC060C1D	...	59,500	7.65	60,000	2.55	35,000	1.90	6.20

GENERAL ELECTRIC COMPANY (CONT.)

Type: HRCU-A-C Trade Name: Weathertron

	Model		Cooling		Heating				
footnote(s)	outdoor unit	indoor unit	capacity (Btuh)	SEER	capacity (Btuh)	COP @ 47°	capacity (Btuh)	COP @ 17°	HSPF
.	BWB712A	BXA718A-HP	14,100	7.45	16,100	2.50	9,300	1.70	5.60
.	BWB712A	BXA724A-HP	14,200	7.50	16,200	2.50	9,300	1.75	5.55
.	BWB712A	BXF724A-HP	14,300	7.55	16,300	2.55	9,300	1.75	5.70
.	BWR718A	BXA718A-HP	15,800	7.95	16,500	2.55	9,000	1.70	5.75
.	BWR718A	BXA724A-HP	15,900	8.10	16,500	2.55	9,000	1.70	5.80
.	BWR718A	BXF724A-HP	16,300	8.25	16,700	2.60	9,100	1.70	5.85
2	BWB718A-B	BXA718A-HP	17,500	7.70	17,900	2.55	9,900	1.70	5.75
2	BWB718A-B	BXA724A-HP	17,900	7.70	18,200	2.60	10,000	1.70	5.80
2	BWB718A-B	BXF724A-HP	18,300	7.75	18,400	2.65	10,100	1.75	5.85
2	BWB724A-B	BXA724A-HP	22,000	7.65	21,400	2.45	12,300	1.70	5.65
2	BWB724A-B	BXF724A-HP	22,600	7.90	21,400	2.50	12,300	1.70	5.70
2	BWB724A-B	BXA730A-HP	23,400	8.00	21,800	2.55	12,500	1.75	5.75
.	BWR724A	BXA724A-HP	25,800	7.85	26,600	2.55	15,000	1.95	6.05
.	BWR724A	BXF724A-HP	26,400	8.15	26,800	2.60	15,000	1.95	6.15
.	BWR724A	BXA730A-HP	28,000	8.40	27,600	2.70	15,400	2.00	6.25
2	BWR724A	BXA748A-HP	28,000	8.40	27,600	2.70	15,400	2.00	6.25

GENERAL ELECTRIC COMPANY (CONT.)

Type: HRCU-A-C Trade Name: Weathertron

	Model		Cooling			Heating			
footnote(s)	outdoor unit	indoor unit	capacity (Btuh)	SEER	capacity (Btuh)	COP @ 47°	capacity (Btuh)	COP @ 17°	HSPF
2	BWB730A	BXA730A-HP	28,000	7.25	28,800	2.55	16,700	1.85	6.05
2	BWB730A	BXA736A -HP	28,000	7.40	29,000	2.55	16,800	1.85	6.05
2	BWB730A	BXF736A-HP	28,600	7.45	29,000	2.60	16,800	1.85	6.10
.	BWR730A	BXF 736A-HP	31,000	8.05	33,400	2.80	18,600	2.00	6.40
.	BWB736A	BXA736A-HP	31,400	6.75	33,200	2.35	20,000	1.75	5.65
2	BWR730A	BXA742A-HP	31,800	8.15	33,600	2.80	18,700	2.00	6.45
.	BWB736A	BXF736A-HP	32,000	6.85	33,400	2.35	20,000	1.75	5.70
.	BWR730A	BXA748-HP	32,600	8.50	33,600	2.85	18,800	2.00	6.55
.	BWB736A	BXA742A-HP	32,800	7.05	33,600	2.40	20,200	1.80	5.75
.	BWR730A	BXF760A-HP	33,400	8.60	34,000	2.90	18,900	2.05	6.60
.	BWR730A	BXA760A-HP	33,600	8.60	34,000	2.90	18,900	2.05	6.65
2	BWB736A	BXA748A-HP	33,600	7.40	33,600	2.40	20,200	1.80	5.90
.	BWR736A	BXF736A-HP	35,000	7.65	37,400	2.65	22,000	2.00	6.20
.	BWR736A	BXA742A-HP	36,200	7.85	38,000	2.70	22,400	2.05	6.35
2	BWB742A-B	BXA742A-HP	37,000	7.15	41,500	2.55	23,400	1.85	5.85
.	BWR736A	BXA748A-HP	37,200	8.20	38,500	2.75	22,600	2.05	6.45

GENERAL ELECTRIC COMPANY (CONT.)

Type: HRCU-A-C Trade Name: Weathertron

	Model		Cooling		Heating				
footnote(s)	outdoor unit	indoor unit	capacity (Btuh)	SEER	capacity (Btuh)	COP @ 47°	capacity (Btuh)	COP @ 17°	HSPF
2	BWB742A-B	BXA748A-HP	37,600	7.55	42,000	2.60	23,600	1.85	5.95
.	BWR736A	BXF760A-HP	38,000	8.30	38,500	2.75	22,600	2.05	6.50
.	BWR742A	BXA742A-HP	38,000	7.45	42,000	2.65	24,600	2.00	6.15
.	BWR736A	BXA760A-HP	38,500	8.30	38,500	2.80	22,600	2.05	6.55
2	BWB742A-B	BXF760A -HP	38,500	7.65	42,500	2.65	23,800	1.85	6.00
.	BWR742A	BXA748A-HP	39,000	8.05	42,500	2.70	24,600	2.00	6.30
2	BWB742A-B	BXA760A-HP	39,000	7.65	42,500	2.65	23,800	1.85	6.05
.	BWR742A	BXF760A-HP	40,000	8.20	43,000	2.75	24,800	2.00	6.35
.	BWR742A	BXA760A-HP	40,500	8.25	43,000	2.75	24,800	2.00	6.40
2	BWB748A-B	BXA748A-HP	42,000	7.80	46,500	2.60	26,800	1.95	6.30
.	BWR748A	BXA748A-HP	42,500	7.95	47,500	2.60	28,400	2.00	6.30
2	BWB748A-B	BXF760A-HP	43,000	7.95	47,000	2.65	27,000	1.95	6.40
2	BWB748A-B	BXA760A-HP	43,500	8.00	47,000	2.65	27,000	1.95	6.40
.	BWR748A	BXF760A-HP	44,000	8.05	48,000	2.65	28,600	2.00	6.40

GENERAL ELECTRIC COMPANY (CONT.)

Type: HRCU-A-C Trade Name: Weathertron

	Model		Cooling			Heating			
footnote(s)	outdoor unit	indoor unit	capacity (Btuh)	SEER	capacity (Btuh)	COP @ 47°	capacity (Btuh)	COP @ 17°	HSPF
.	BWR748A	BXA760A-HP	44,500	8.10	48,000	2.65	28,600	2.05	6.40
.	BWR760A	BXF760A-HP	54,000	7.20	60,500	2.60	36,400	2.00	6.35
.	BWR760A	BXA760A-HP	54,500	7.25	61,000	2.60	36,400	2.05	6.35

Type: HRCU-A-CB Trade Name: Weathertron

footnote(s)	outdoor unit	indoor unit	capacity (Btuh)	SEER	capacity (Btuh)	COP @ 47°	capacity (Btuh)	COP @ 17°	HSPF
.	BWB712A	BWU018A	14,000	7.50	16,200	2.50	9,300	1.75	5.65
.	BWB712A	BWH018A	14,300	7.75	16,200	2.50	9,300	1.70	5.65
.	BWB712A	BWV018A	14,400	7.85	16,100	2.50	9,200	1.75	5.65
.	BWB712A	BWH718A	14,900	7.75	16,500	2.60	9,400	1.75	5.75
.	BWB712A	* BWV718A	15,000	7.85	16,400	2.60	9,400	1.75	5.80
.	BWR718A	BWV018A	16,300	7.85	17,900	2.70	10,100	1.95	6.50
.	BWR718A	BWH018A	16,300	7.75	18,000	2.70	10,100	1.95	6.45
.	BWR718A	BWU018A	16,300	7.60	18,000	2.70	10,100	1.95	6.50
.	BWR718A	BWU024A	16,600	7.70	18,200	2.75	10,200	1.95	6.60

GENERAL ELECTRIC COMPANY (CONT.)

Type: HRCU-A-CB Trade Name: Weathertron

	Model		Cooling			Heating			
footnote(s)	outdoor unit	indoor unit	capacity (Btuh)	SEER	capacity (Btuh)	COP @ 47°	capacity (Btuh)	COP @ 17°	HSPF
.	BWR718A	BWH718A	17,000	7.65	18,300	2.80	10,200	1.95	6.65
.	BWR718A	BWH724A	17,000	7.95	18,300	2.85	10,200	2.00	6.70
.	BWR718A	BWV724A	17,000	7.90	18,400	2.85	10,200	2.00	6.70
.	BWR718A	BEV024A/BXV724A	17,000	7.90	18,400	2.85	10,200	2.00	6.70
.	BWR718A	BWV718A	17,100	7.75	18,300	2.80	10,200	2.00	6.65
.	BWR718A	BWV024A	17,200	8.00	18,400	2.80	10,300	2.00	6.65
.	BWR718A	BWH024A	17,300	8.15	18,300	2.85	10,200	2.00	6.70
.	BWV718A-B	BWV018A	17,600	8.15	17,900	2.55	9,800	1.70	5.85
.	BWB718A-B	BWH018A	17,600	8.05	18,000	2.55	9,900	1.70	5.85
2	BWB718A-B	BWU018A	17,600	7.70	18,000	2.55	9,900	1.70	5.80
.	BWR718A	BWV030A	17,800	8.05	18,700	2.85	10,400	2.00	6.70
.	BWR718A	BWH030A	17,900	8.10	18,700	2.90	10,400	2.00	6.75
.	BWB718A-B	BWU024A	18,100	7.75	18,400	2.65	10,100	1.75	5.90
.	BWB718A-B	BWH718A	18,400	7.65	18,300	2.65	10,000	1.75	5.90
2	BWB718A-B	* BWV718A	18,500	7.75	18,200	2.65	10,000	1.75	5.90

GENERAL ELECTRIC COMPANY (CONT.)

Type: HRCU-A-CB Trade Name: Weathertron

	Model		Cooling			Heating			
footnote(s)	outdoor unit	indoor unit	capacity (Btuh)	SEER	capacity (Btuh)	COP @ 47°	capacity (Btuh)	COP @ 17°	HSPF
.	BWB718A-B	BWV724A	18,500	7.90	18,500	2.70	10,100	1.75	5.95
.	BWB718A-B	BWH724A	18,500	7.95	18,400	2.70	10,100	1.75	5.95
.	BWB718A-B	BEV024A/BXV724A	18,500	7.90	18,500	2.70	10,100	1.75	5.95
.	BWB718A-B	BWV024A	18,700	8.35	18,500	2.70	10,200	1.75	6.00
.	BWB718A-B	BWH024A	18,800	8.45	18,400	2.70	10,100	1.75	6.05
.	BWR718A	BWH736S	19,200	8.65	19,000	3.10	10,400	2.10	7.15
.	BWR718A	BWV736S	19,200	8.65	19,000	3.10	10,400	2.10	7.10
.	BWB718A-B	BWV030A	19,400	8.35	18,700	2.70	10,300	1.75	6.05
.	BWB718A-B	BWH030A	19,500	8.45	18,700	2.70	10,200	1.75	6.05
.	BWB724A-B	BWU024A	23,400	8.00	21,400	2.50	12,400	1.75	5.70
.	BWB724A-B	BWV024A	23,800	8.55	21,600	2.55	12,400	1.75	5.85
.	BWB724A-B	BWU030A	23,800	8.15	21,600	2.55	12,400	1.75	5.75
.	BWB724A-B	BWH024A	24,000	8.70	21,400	2.55	12,300	1.75	5.90
.	BWB724A-B	BWH724A	24,000	8.05	21,400	2.55	12,300	1.75	5.80

GENERAL ELECTRIC COMPANY (CONT.)

Type: HRCU-A-CB Trade Name: Weathertron

	Model			Cooling			Heating		
footnote(s)	outdoor unit	indoor unit	capacity (Btuh)	SEER	capacity (Btuh)	COP @ 47°	capacity (Btuh)	COP @ 17°	HSPF
.	BWB724A-B	* BWV724A	24,000	8.00	21,600	2.55	12,300	1.75	5.75
.	BWB724A-B	BEV024A/BXB724A	24,000	8.00	21,600	2.55	12,300	1.75	5.75
.	BWB724A-B	BWV730A	24,800	8.15	22,000	2.60	12,700	1.75	5.80
.	BWB724A-B	BWV030A	25,000	8.55	22,000	2.55	12,700	1.75	5.90
.	BWB724A-B	BWH730A	25,000	8.20	22,000	2.60	12,700	1.75	5.85
.	BWB724A-B	BWH030A	25,200	8.60	22,000	2.60	12,700	1.75	5.95
.	BWB724A-B	BWV736S	26,800	8.90	22,000	2.75	12,600	1.85	6.10
.	BWB724A-B	BWH736S	27,000	8.90	22,000	2.75	12,600	1.85	6.10
.	BWR724A	BWH024A	27,600	8.70	26,200	2.65	14,000	1.80	6.00
.	BWR724A	BWH724A	27,600	8.15	26,200	2.65	14,000	1.80	5.90
.	BWR724A	* BWV724A	27,600	8.10	26,400	2.65	14,100	1.80	5.90
.	BWR724A	BEV024A/BXV724A	27,600	8.10	26,400	2.65	14,100	1.80	5.90
.	BWR724A	BWV024A	27,600	8.60	26,400	2.60	14,100	1.80	5.95
.	BWR724A	BWU030A	27,600	8.25	26,600	2.65	14,200	1.80	5.90
.	BWR724A	BWV034F	27,600	8.25	26,600	2.65	14,200	1.80	5.90

GENERAL ELECTRIC COMPANY (CONT.)

Type: HRCU-A-CB Trade Name: Weathertron

	Model		Cooling			Heating			
footnote(s)	outdoor unit	indoor unit	capacity (Btuh)	SEER	capacity (Btuh)	COP @ 47°	capacity (Btuh)	COP @ 17°	HSPF
.	BWB730A	BWU030A	28,200	7.35	29,600	2.55	16,500	1.85§	5.95§
.	BWR724A	BWH730A	29,000	8.40	27,200	2.70	14,500	1.85	6.05
.	BWR724A	BWV730A	29,000	8.40	27,200	2.70	14,500	1.85	6.00
.	BWR724A	BWV030A	29,200	8.70	27,200	2.70	14,500	1.85	6.05
.	BWR724A	BWH030A	29,400	8.75	27,200	2.70	14,500	1.85	6.10
.	BWB730A	* BWV730A	29,600	7.60	30,200	2.60	16,900§	1.85§	6.05§
.	BWB730A	BWV030A	29,600	8.05	30,200	2.60	16,900§	1.85§	6.15§
.	BWB730A	BWH030A	29,800	8.15	30,000	2.60	16,800§	1.90§	6.15§
.	BWB730A	BWH730A	29,800	7.60	30,000	2.60	16,800§	1.90§	6.05§
.	BWB730A	BWH736A	31,000	7.90	30,600	2.70	17,200§	1.90§	6.15§
.	BWB730A	BWH036A	31,000	7.90	30,600	2.70	17,200§	1.90§	6.15§
.	BWB730A	BWV036A	31,000	7.90	30,600	2.70	17,200§	1.90§	6.15§
.	BWB730A	BWV736A	31,000	7.90	30,600	2.70	17,200§	1.90§	6.15§
.	BWB730A	BEV036A/BXV736A	31,000	7.90	30,600	2.70	17,200§	1.90§	6.15§
.	BWR730A	* BWV730A	31,400	8.00	32,600	2.70	17,200	1.90	6.05

GENERAL ELECTRIC COMPANY (CONT.)

Type: HRCU-A-CB Trade Name: Weathertron

	Model		Cooling		Heating				
footnote(s)	outdoor unit	indoor unit	capacity (Btuh)	SEER	capacity (Btuh)	COP @ 47°	capacity (Btuh)	COP @ 17°	HSPF
·	BWR730A	BWV030A	31,400	8.45	32,600	2.70	17,300	1.90	6.10
·	BWR724A	BWH736S	31,600	9.20	27,600	2.95	14,500	1.95	6.35
·	BWR724A	BWV736S	31,600	9.20	27,600	2.95	14,500	1.95	6.35
·	BWR730A	BWH030A	31,600	8.55	32,600	2.70	17,200	1.90	6.10
·	BWR730A	BWH730A	31,600	8.05	32,600	2.70	17,200	1.90	6.05
·	BWB730A	BWH042A	31,600	8.20	30,600	2.70	17,200§	1.90§	6.25§
·	BWB730A	BWV042A	31,800	8.25	30,600	2.70	17,200§	1.90§	6.25§
·	BWB730A	BWV034F	31,800	7.90	30,800	2.75	17,300§	1.90§	6.25§
·	BWB730A	BWH736S	32,600	8.50	30,600	2.85	17,100§	2.00§	6.45§
·	BWB730A	BWV736S	32,600	8.50	30,600	2.85	17,100§	2.00§	6.45§
·	BWR730A	BWH036A	33,000	8.40	33,400	2.80	17,600	1.95	6.20
·	BWR730A	BWH736A	33,000	8.40	33,400	2.80	17,600	1.95	6.20
·	BWR730A	BWV036A	33,000	8.40	33,400	2.80	17,600	1.95	6.20
·	BWR730A	BWV736A	33,000	8.40	33,400	2.80	17,600	1.95	6.20
·	BWR730A	BEV036A/BXV736A	33,000	8.40	33,400	2.80	17,600	1.95	6.20

GENERAL ELECTRIC COMPANY (CONT.)

Type: HRCU-A-CB Trade Name: Weathertron

	Model		Cooling			Heating			
footnote(s)	outdoor unit	indoor unit	capacity (Btuh)	SEER	capacity (Btuh)	COP @ 47°	capacity (Btuh)	COP @ 17°	HSPF
.	BWS736A	* BWV736S	34,000	11.25	34,000	2.95	18,700	2.00	7.10
.	BWR730A	BWH042A	34,000	8.75	33,600	2.80	17,700	1.95	6.25
.	BWS736A	BWH736S	34,200	11.25	34,000	3.00	18,700	2.00	7.10
.	BWR730A	BWV042A	34,200	8.80	33,600	2.85	17,700	1.95	6.30
.	BWR730A	BWV034F	34,200	8.50	33,800	2.85	17,800	1.95	6.25
.	BWB736A	BWH736A	34,800	7.25	34,400	2.55	20,200	1.85	6.00
.	BWB736A	BWV036A	34,800	7.25	34,400	2.55	20,200	1.85	6.00
.	BWB736A	BWV736A	34,800	7.25	34,400	2.55	20,200	1.85	6.00
.	BWB736A	* BEV036A/BXV736A	34,800	7.25	34,400	2.55	20,200	1.85	6.00
.	BWB736A	BWH036A	34,800	7.25	34,400	2.55	20,200	1.85	6.00
.	BWR730A	BWV736S	35,000	9.10	33,800	2.95	17,600	2.05	6.50
.	BWR730A	BWH736S	35,200	9.15	33,800	3.00	17,600	2.05	6.50
.	BWB736A	BWH742A	35,800	7.45	34,800	2.60	20,400	1.85	6.05
.	BWB736A	BWH042A	36,000	7.85	35,000	2.60	20,600	1.85	6.20
.	BWB736A	BWV742A	36,000	7.50	34,800	2.60	20,400	1.85	6.10

GENERAL ELECTRIC COMPANY (CONT.)

Type: HRCU-A-CB Trade Name: Weathertron

	Model		Cooling			Heating			
footnote(s)	outdoor unit	indoor unit	capacity (Btuh)	SEER	capacity (Btuh)	COP @ 47°	capacity (Btuh)	COP @ 17°	HSPF
.	BWB736A	BWV042A	36,200	7.90	34,800	2.60	20,400	1.85	6.20
.	BWB736A	BWV034F	36,400	7.35	34,600	2.60	20,400	1.85	6.05
.	BWB736A	BWH736S	36,800	8.15	34,600	2.70	20,000	1.90	6.35
.	BWB736A	BWV736S	36,800	8.10	34,600	2.70	20,000	1.90	6.30
.	BWR736A	BWH036A	37,400	8.00	38,500	2.65	21,200	1.90	6.20
.	BWR736A	BWH736A	37,400	8.00	38,500	2.65	21,200	1.90	6.20
.	BWR736A	BWV036A	37,400	8.00	38,500	2.65	21,200	1.90	6.20
.	BWR736A	* BWV736A	36,400	8.00	38,500	2.65	21,200	1.90	6.20
.	BWR736A	BEV036A/BXV736A	37,400	8.00	38,500	2.65	21,200	1.90	6.20
.	BWR736A	BWH742A	38,500	8.25	39,500	2.70	21,600	1.90	6.30
.	BWR736A	BWV034F	39,000	8.15	39,000	2.70	21,400	1.90	6.25
.	BWR736A	BWH042A	39,000	8.50	39,500	2.70	21,600	1.90	6.30
.	BWR736A	BWV042A	39,000	8.55	39,000	2.75	21,600	1.90	6.35
.	BWR736A	BWV742A	39,000	8.30	39,000	2.75	21,600	1.90	6.30
.	BWR736A	BWV736S	39,500	8.85	39,000	2.85	21,200	2.00	6.50

GENERAL ELECTRIC COMPANY (CONT.)

Type: HRCU-A-CB Trade Name: Weathertron

	Model		Cooling			Heating			
footnote(s)	outdoor unit	indoor unit	capacity (Btuh)	SEER	capacity (Btuh)	COP @ 47°	capacity (Btuh)	COP @ 17°	HSPF
.	BWR736A	BWH736S	39,500	8.90	39,000	2.85	21,200	2.00	6.50
.	BWR742A	BWE060C	41,000	8.15	43,000	2.80	25,000	2.05	6.35
.	BWB742A-B	BWV742A	42,000	7.80	42,500	2.65	23,800	1.85	6.00
.	BWB742A-B	BWV042A	42,000	8.30	42,500	2.65	23,800	1.85	6.05
.	BWB742A-B	BWH042A	42,000	8.25	42,500	2.65	24,000	1.85	6.05
.	BWB742A-B	BWH742A	42,000	7.80	42,500	2.65	24,000	1.85	6.00
.	BWS742A	* BWV760S	42,000	10.00	44,000	2.90	24,600	1.95	6.85
.	BWS742A	BWH760S	42,500	10.00	44,000	2.90	24,600	1.95	6.85
.	BWR742A	BWV042A	43,000	8.65	47,000	2.80	27,400	2.15	6.75
.	BWR742A	* BWV742A	43,000	8.15	47,000	2.80	27,400	2.15	6.70
.	BWR742A	BWH042A	43,000	8.60	47,000	2.80	27,600	2.10	6.75
.	BWR742A	BWH742A	43,000	8.15	47,000	2.80	27,600	2.10	6.70
.	BWB742A-B	BWH748A	44,500	8.15	43,500	2.75	24,400	1.90	6.15
.	BWB742A-B	BWV748A	44,500	8.05	44,000	2.75	24,800	1.90	6.10
2	BWB748A-B	BWE060C	44,500	7.95	47,500	2.70	27,200	1.95	6.40

GENERAL ELECTRIC COMPANY (CONT.)

Type: HRCU-A-CB Trade Name: Weathertron

	Model		Cooling			Heating			
footnote(s)	outdoor unit	indoor unit	capacity (Btuh)	SEER	capacity (Btuh)	COP @ 47°	capacity (Btuh)	COP @ 17°	HSPF
·	BWR748A	BWE060C	45,000	8.00	48,500	2.70	28,800	2.05	6.40
·	BWR742A	BWH748A	46,000	8.65	48,500	2.95	28,000	2.20	6.90
·	BWR742A	BWV748A	46,000	8.50	48,500	2.95	28,200	2.15	6.85
·	BWB742A-B	BWH760A	46,000	8.20	44,500	2.80	25,000	1.90	6.15
·	BWB742A-B	BWV760A	46,000	8.15	44,500	2.75	25,200	1.90	6.15
·	BWB742A-B	BWH760 S	46,500	8.45	44,500	2.90	24,800	1.95	6.35
·	BWB742A-B	BWV760S	46,500	8.45	44,500	2.90	24,800	1.95	6.35
·	BWR742A	BWH760A	48,000	8.70	49,500	3.00	28,600	2.15	6.95
·	BWR742A	BWV760A	48,000	8.60	49,500	2.95	28,800	2.15	6.90
·	BWR742A	BWV760S	48,500	8.95	49,500	3.10	28,400	2.25	7.15
·	BWR742A	BWH760S	48,500	8.95	49,500	3.10	28,400	2.25	7.15
·	BWB748A-B	BWH748A	49,000	8.10	47,500	2.75	27,200	2.00	6.45
·	BWB748A-B	* BWV748A	49,000	8.00	48,000	2.75	27,600	1.95	6.45
·	BWR748A	BWH748A	49,500	8.40	51,500	2.90	30,200	2.20	6.90

GENERAL ELECTRIC COMPANY (CONT.)

Type: HRCU-A-CB Trade Name: Weathertron

footnote(s)	Model: outdoor unit	Model: indoor unit	Cooling: capacity (Btuh)	Cooling: SEER	Heating: capacity (Btuh)	Heating: COP @ 47°	Heating: capacity (Btuh)	Heating: COP @ 17°	HSPF
.	BWR748A	* BWV748A	49,500	8.30	51,500	2.85	30,400	2.15	6.85
.	BWB748A-B	BWV760A	51,500	8.35	49,500	2.80	28,400	1.95	6.50
.	BWB748A-B	BWV760S	52,000	8.65	49,000	2.90	28,200	2.05	6.65
.	BWB748A-B	BWH760 S	52,000	8.65	49,000	2.90	28,200	2.05	6.65
.	BWB748A-B	BWH760A	52,000	8.45	49,000	2.80	28,400	2.00	6.50
.	BWR748A	BWV760A	52,000	8.60	53,000	2.90	31,400	2.15	6.95
.	BWR748A	BWH760S	52,500	8.90	53,000	3.05	31,000	2.25	7.20
.	BWR748A	BWV760S	52,500	8.90	53,000	3.05	31,000	2.25	7.15
.	BWR748A	BWH760A	52,500	8.70	53,000	2.95	31,200	2.20	7.00
.	BWS754A	* BWV760S	53,500	9.55	56,500	2.90	32,400	1.95	6.75
.	BWS754A	BWH760S	54,000	9.55	56,500	2.90	32,400	1.95	6.75
.	BWR760A	BWE060C	55,500	7.15	61,000	2.65	36,600	2.05	6.40
.	BWR760A	BWH760A	59,000	7.30	62,500	2.75	37,400	2.05	6.55
.	BWR760A	* BWV760A	59,000	7.25	62,500	2.75	37,600	2.05	6.50
.	BWR760A	BWH760S	59,500	7.45	62,500	2.85	37,200	2.10	6.65
.	BWR760A	BWV760S	59,500	7.45	62,500	2.85	37,200	2.10	6.65

GENERAL ELECTRIC COMPANY (CONT.)
Products Not Covered by DOE

Type: HSP-A Trade Name: Weathertron

	Model		Cooling		Heating				
footnote(s)	outdoor unit	indoor unit	capacity (Btuh)	EER	capacity (Btuh)	COP @ 47°	capacity (Btuh)	COP @ 17°	HSPF
.	BWCO36B3	...	37,000	6.75	40,000	2.35	20,500	1.50	...
.	BWCO36B4	...	37,000	6.75	40,000	2.35	20,500	1.50	...
.	BWCO36E3	...	37,800	8.20	41,000	2.95	21,400	1.95	...
.	BWCO36E4	...	37,800	8.20	41,000	2.95	21,400	1.95	...
.	BWCO42E3	...	41,000	6.70	50,000	2.70	27,000	1.90	...
.	BWCO42E4	...	41,000	6,70	50,000	2.70	27,000	1.90	...
1	BWCO42C3	...	43,000	7.50	46,500	2.65	26,400	1.80	...
.	BWCO42C4	...	43,000	7.50	46,500	2.65	26,400	1.80	...
1	BWCO48C3	...	48,500	7.50	48,500	2.60	28,600	1.85	...
.	BWCO48C4	...	48,500	7.50	48,500	2.60	28,600	1.85	...
1	BWCO60C3G	...	55,500	7.85	55,000	2.75	32,200	1.95	...
.	BWCO60C4G	...	55,500	7.85	55,000	2.75	32,200	1.95	...
1	BWCO60C3D	...	59,500	7.10	60,000	2.55	35,000	1.90	...
.	BWCO60C4D	...	59,500	7.10	60,000	2.55	35,000	1.90	...
.	BWCO90C2	...	90,000	7.85	90,000	2.65	46,000§	1.75§	...

GENERAL ELECTRIC COMPANY (CONT.)

Type: HSP-A Trade Name: Weathertron

	Model		Cooling		Heating			
footnote(s)	outdoor unit	indoor unit	capacity (Btuh)	EER	capacity (Btuh)	COP @ 47°	capacity (Btuh)	COP @ 17°	HSPF
·	BWCO90C3	...	90,000	7.85	90,000	2.65	46,000§	1.75§	...
·	BWCO90C4	...	90,000	7.85	90,000	2.65	46,000§	1.75§	...
·	BWCO90F2	...	94,000	8.70	88,000	2.70	48,000	1.90	...
·	BWCO90F3	...	94,000	8.70	88,000	2.70	48,000	1.90	...
·	BWCO90F4	...	94,000	8.70	88,000	2.70	48,000	1.90	...
·	BWC120F2	...	116,000	7.80	112,000	2.75	58,000	1.85	...
·	BWC120F3	...	116,000	7.80	112,000	2.75	58,000	1.85	...
·	BWC120F4	...	116,000	7.80	112,000	2.75	58,000	1.85	...
·	BWC120C2	...	117,000	8.00	112,000	2.75	58,000	1.75§	...
·	BWC120C3	...	117,000	8.00	112,000	2.75	58,000	1.75§	...
·	BWC120C4	...	117,000	8.00	112,000	2.75	58,000	1.75§	...

Type: HRCU-A-C Trade Name: Weathertron

footnote(s)	outdoor unit	indoor unit	capacity (Btuh)	EER	capacity (Btuh)	COP @ 47°	capacity (Btuh)	COP @ 17°	HSPF
	BWR036A	BXF736A-HP	35,000	7.50	37,400	2.65	22,000	2.00	...

GENERAL ELECTRIC COMPANY (CONT.)

Type: HRCU-A-C Trade Name: Weathertron

Model			Cooling		Heating				
footnote(s)	outdoor unit	indoor unit	capacity (Btuh)	EER	capacity (Btuh)	COP @ 47°	capacity (Btuh)	COP @ 17°	HSPF
.	BWR036A	BXA742A-HP	36,200	7.50	38,000	2.70	22,400	2.05	...
.	BWR036A	BXA748A-HP	37,200	7.70	38,500	2.75	22,600	2.05	...
.	BWR036A	BXF760A-HP	38,000	7.80	38,500	2.75	22,600	2.05	...
.	BWR036A	BXA760A-HP	38,500	7.80	38,500	2.80	22,600	2.05	...
.	BWR048A	BXA748A-HP	42,500	7.10	47,500	2.60	28,400	2.00	...
.	BWR048A	BXF760A-HP	44,000	7.20	48,000	2.65	28,600	2.00	...
.	BWR048A	BXA760A-HP	44,500	7.25	48,000	2.65	28,600	2.05	...
.	BWR060A	BXF760A-HP	54,000	6.70	60,500	2.60	36,400	2.00	...
.	BWR060A	BXA760A-HP	54,500	6.70	61,000	2.60	36,400	2.05	...
.	BWA090C2	BGXA120A	95,000§	8.00§	92,000§	2.75§	47,000§	1.75§	...
.	BWA090C3	BGXA120A	95,000§	8.00§	92,000	2.75§	47,000§	1.75§	...
.	BWA090C4	BGXA120A	95,000§	8.00§	92,000	2.75§	47,000§	1.75§	...
.	BWA120C2	BGXA120A	117,000	8.00§	112,000	2.75	58,000	1.75§	...
.	BWA120C3	BGXA120A	117,000	8.00§	112,000	2.75	58,000	1.75§	...
.	BWA120C4	BGXA120A	117,000	8.00§	112,000	2.75	58,000	1.75§	...

GENERAL ELECTRIC COMPANY (CONT.)

Type: HRCU-A-CB Trade Name: Weathertron

	Model			Cooling			Heating		
footnote(s)	outdoor unit	indoor unit	capacity (Btuh)	EER	capacity (Btuh)	COP @ 47°	capacity (Btuh)	COP @ 17°	HSPF
.	BWR036A	BWH036A	37,400	7.75	38,500	2.65	21,200	1.90	...
.	BWR036A	BWH736A	37,400	7.75	38,500	2.65	21,200	1.90	...
.	BWR036A	BWV036A	37,400	7.75	38,500	2.65	21,200	1.90	...
.	BWR036A	BWV736A	37,400	7.75	38,500	2.65	21,200	1.90	...
.	BWR036A	BEV036A/BXV736A	37,400	7.75	38,500	2.65	21,200	1.90	...
.	BWR036A	BWE060C	38,500	7.85	38,500	2.80	22,800	2.05	...
.	BWR036A	BWH742A	38,500	7.70	39,500	2.70	21,600	1.90	...
.	BWR036A	BWV042A	39,000	7.85	39,000	2.75	21,600	1.90	...
.	BWR036A	BWHO42A	39,000	7.80	39,500	2.70	21,600	1.90	...
.	BWR036A	BWV742A	39,000	7.80	39,000	2.75	21,600	1.90	...
.	BWR036A	BWH736S	39,500	8.20	39,000	2.85	21,200	2.00	...
.	BWR036A	BWV736S	39,500	8.15	39,000	2.85	21,200	2.00	...
.	BWR048A	BWE060C	45,000	7.30	48,500	2.70	28,800	2.05	...
.	BWR048A	BWH748A	49,500	7.95	51,500	2.90	30,200	2.20	...
.	BWR048A	BWV748A	49,500	7.80	51,500	2.85	30,400	2.15	...

GENERAL ELECTRIC COMPANY (CONT.)

Type: HRCU-A-CB Trade Name: Weathertron

	Model		Cooling		Heating				
footnote(s)	outdoor unit	indoor unit	capacity (Btuh)	EER	capacity (Btuh)	COP @ 47°	capacity (Btuh)	COP @ 17°	HSPF
.	BWR048A	BWV760A	52,000	7.80	53,000	2.90	31,400	2.15	...
.	BWR048A	BWH760S	52,500	8.05	53,000	3.05	31,000	2.25	...
.	BWR048A	BWV760S	52,500	8.05	53,000	3.05	31,000	2.25	...
.	BWR048A	BWH760A	52,500	7.85	53,000	2.95	31,200	2.20	...
.	BWR060A	BWE060C	55,500	6.80	61,000	2.65	36,600	2.05	...
.	BWR060A	BWH760A	59,000	6.85	62,500	2.75	37,400	2.05	...
.	BWR060A	BWV760A	59,000	6.80	62,500	2.75	37,600	2.05	...
.	BWR060A	BWH760S	59,500	7.00	62,500	2.85	37,200	2.10	...
.	BWR060A	BWV760S	59,500	7.00	62,500	2.85	37,200	2.10	...
.	BWR060A	BWE090C	62,500	7.15	62,500	2.85	37,400	2.10	...
.	BWA090C2	BWE090C	90,000	7.85	90,000	2.65	46,000§	1.75§	...
.	BWA090C3	BWE090C	90,000	7.85	90,000	2.65	46,000§	1.75§	...
.	BWA090C4	BWE090C	90,000	7.85	90,000	2.65	46,000§	1.75§	...
.	BWA090C2	BWE120C	95,000	8.00§	90,000§	2.75§	47,000§	1.75§	...
.	BWA090C3	BWE120C	95,000	8.00§	90,000§	2.75§	47,000§	1.75§	...

GENERAL ELECTRIC COMPANY (CONT.)

Type: HRCU-A-CB Trade Name: Weathertron

	Model		Cooling		Heating				
footnote(s)	outdoor unit	indoor unit	capacity (Btuh)	SEER	capacity (Btuh)	COP @ 47°	capacity (Btuh)	COP @ 17°	HSPF
.	BWA120C2	BWE120C	117,000	8.00	112,000	2.75	58,000	1.75§	...
.	BWA120C3	BWE120C	117,000	8.00	112,000	2.75	58,000	1.75§	...
.	BWA120C4	BWE120C	117,000	8.00	112,000	2.75	58,000	1.75§	...

LENNOX INDUSTRIES, INC.
DOE Covered Products

Type: HSP-A Trade Name: Lennox

	Model		Cooling		Heating				
footnote(s)	outdoor unit	indoor unit	capacity (Btuh)	EER	capacity (Btuh)	COP @ 47°	capacity (Btuh)	COP @ 17°	HSPF
.	CHP9-261-2P	...	22,800	6.40	26,200	2.40	15,400	1.65	5.80
1	CHP10-261-1P	...	24,800	7.95	27,400	2.60	13,800	1.55	5.45
.	CHP9-311-2P	...	27,800	6.50	31,200	2.40	17,800	1.60	5.60
1	CHP10-311-1P	...	28,800§	8.05§	31,000§	2.50§	17,900§	§1.70	6.75§
.	CHP9-411-4P	...	34,200	6.25	38,500	2.35	22,200	1.60	5.75
.	CHP10-411-2P	...	35,000	7.80	35,800	2.65	21,400	1.90	6.35
2	CHP10-461-1P	...	40,000	8.05	42,500	2.75	23,200	1.85	6.55
.	CHP9-461-3P	...	41,000	6.65	42,500	2.30	26,800	1.65	5.50
2	CHP10-511-1P	...	46,000	7.70	47,000	2.55	27,200	1.75	5.95
.	CHP9-511-3P	...	46,000	6.40	50,000	2.30	30,000	1.45	5.60
.	CHP10-651-1P	...	54,000	6.55	60,000	2.50	36,400	1.80	5.85
2	CHP10B-651-1P	...	55,000	7.70	57,000	2.70	32,400	1.85	6.20

LENNOX INDUSTRIES, INC. (CONT.)

Type: HRCU-A-C Trade Name: Lennox

	Model		Cooling		Heating				
footnote(s)	outdoor unit	indoor unit	capacity (Btuh)	SEER	capacity (Btuh)	COP @ 47°	capacity (Btuh)	COP @ 17°	HSPF
.	HP 16-211V-3A,11A	* CP12-26-1	18,700	9.30	20,600	2.95	11,000	2.00	7.20
2	HP 10-211V-1A,2A	* C12-420-1	18,700	8.35	19,200	2.70	10,500	1.80	6.60
.	HP 10-211V-1A,2A	CR12-420-1	18,700	8.35	19,200	2.70	10,500	1.80	6.60
2	HP 10-261V-1A,2A	* C12-525-1	22,800	8.25	24,400	2.80	14,500	2.10	7.20
2	HP 10-261V-1A,2A	CR12-525-1	22,800	8.25	24,400	2.80	14,500	2.10	7.20
2	HP 10-261V-1A,2A	C12,CR12-420-1	22,800	8.05	24,400	2.85	14,500	2.10	7.25
.	HP 9-261V-3A,4A	* C12-420-1	23,600	7.30	24,800	2.50	13,200	1.75	6.15
.	HP 9-261V-3A,4A	CR12-420-1	23,600	7.30	24,800	2.50	13,200	1.75	6.15
.	HP 16-261V-3A,11A	* CP12-31-1	24,400	9.80	26,000	3.10	14,000	2.05	7.60
.	HP 10-311V-4A,5A	CPH10-490-1	28,800§	7.15§	30,200	2.30	18,200	1.75	5.95§
.	HP 16-311V-3A,11A	* CP12-31-1	29,000	9.20	31,800	3.00	17,900	2.05	7.45
.	HP 10-411V-6A,7A	CPH10-490-1	30,000	7.45	33,200	2.45§	18,400	§1.70	6.60
.	HP 9-311V-3A,4A	* C12-525-1	30,200	7.15	30,000	2.60	18,000	1.90	6.30
.	HP 9-311V-3A,4A	CR12-525-1	30,200	7.15	30,000	2.60	18,000	1.90	6.30
2	HP 10-311V-4A,5A	CR12-525-1	30,800§	7.25	31,400	2.45§	19,200§	1.85	6.35§

LENNOX INDUSTRIES, INC. (CONT.)

Type: HRCU-A-C Trade Name: Lennox

	Model		Cooling		Heating				
footnote(s)	outdoor unit	indoor unit	capacity (Btuh)	SEER	capacity (Btuh)	COP @ 47°	capacity (Btuh)	COP @ 17°	HSPF
2	HP 10-311V-4A,5A	* C12-525-1	31,000§	7.60§	31,800	2.55§	19,400	1.90	6.45§
.	HP 10-411V-6A,7A	* CP12-630-1	32,000	8.10	33,200	2.75	19,300	2.00	6.80
2	HP 10-411V-6A,7A	CR12-630N-1	32,000	8.10	33,200	2.75	19,300	2.00	6.80
.	HP 10-411V-6A,7A	C12-630-1	32,200.	8.05	33,400	2.75	19,400	2.00	6.80
2	HP 10-411V-6A,7A	CR12-630-1	32,200	8.05	33,400	2.75	19,400	2.00	6.80
.	HP 10-311V-4A,5A	CP10-41-1	32,400§	7.35§	31,800	2.55§	19,300	1.85	6.40§
.	HP 10-411V-6A,7A	CP10-41-1	33,400	8.40	33,600	2.75	19,600	2.00	6.80
.	HP 16-411V-3A,11A	* CP12-41-1	35,200	9.30	35,800	2.95	19,600	2.00	7.55
.	HP 9-411V-5A,6A	* CP12-630-1	36,200	7.00	37,200	2.50	21,600	1.85	6.15
.	HP 9-461V-2A	* CP12-630-1	40,500	7.25	42,000	2.50	24,200	1.85	6.20
.	HP 9-461V-2A	C12-630-1	40,500	7.25	42,000	2.50	24,200	1.85	6.20
.	HP 9-461V-2A	CR12-630-1	40,500	7.25	42,000	2.50	24,200	1.85	6.20
.	HP 10-511V-6A,7A	CPH10-645-1	41,500§	7.30§	49,000§	2.70§	30,200§	2.05§	6.95§
.	HP 16-461V-3A,11A	* CP12-461	42,000	9.10	42,000	2.80	25,600	2.10	7.60
2	HP 16-511V-1P,3P	* CP12-51V-1	46,000	8.55	49,500	3.00	29,600§	2.20§	7.55

LENNOX INDUSTRIES, INC. (CONT.)

Type: HRCU-A-C Trade Name: Lennox

	Model		Cooling			Heating			
footnote(s)	outdoor unit	indoor unit	capacity (Btuh)	SEER	capacity (Btuh)	COP @ 47°	capacity (Btuh)	COP @ 17°	HSPF
2	HP 10-511V-6A,7A	CR12-840-1	46,000§	7.70§	48,500§	2.75§	29,600§	2.05§	7.05§
.	HP 10-511V-6A,7A	* C12-840-1	46,500§	8.20§	48,500§	2.75 §	29,400§	2.05§	7.20§
.	HP 10-511V-6A,7A	CP10-51-1	47,500§	7.80§	48,500§	2.75§	29,000§	2.00§	6.90§
.	HP 9-511V-5A,6A	* C12-840-1	47,500	7.05	53,000	2.70	30,800	1.95	6.85
.	HP 9-511V-5A,6A	CR12-840-1	47,500	7.05	53,000	2.70	30,800	1.95	6.85
.	HP 16-651V-1P,3P	* CP12-65V-1	56,000	8.70	61,000	3.00	36,400	2.30	7.80
.	HP 9-651V-2A	* C12-1120-1	56,000	7.55	59,000	2.70	35,800	2.05	6.85
.	HP 9-651V-2A	CR12-1120-1	56,000	7.55	59,000	2.70	35,800	2.05	6.85

Type: HRCU-A-CB Trade Name: Lennox

footnote(s)	outdoor unit	indoor unit	capacity (Btuh)	SEER	capacity (Btuh)	COP @ 47°	capacity (Btuh)	COP @ 17°	HSPF
.	HP 10-211V-1A,2A	CBP13-480-20FF	18,600	8.35	19,200	2.75	10,500	1,85	6.65
.	HP 10-261V-1A,2A	– CBP13-480-20FF	22,600	7.80	24,400	2.80	14,500	2.00	7.10
.	HP 9-261V-3A,4A	CBP13-480FF	24,400	7.50	25,400	2.40	13,700	1.55	5.70
.	HP 9-311V-3A,4A	CBP13-480FF	30,600	7.00	30,400	2.60	19,300	1.95	6.50

LENNOX INDUSTRIES, INC. (CONT.)

Type: HRCU-A-CB Trade Name: Lennox

	Model		Cooling		Heating				
footnote(s)	outdoor unit	indoor unit	capacity (Btuh)	SEER	capacity (Btuh)	COP @ 47°	capacity (Btuh)	COP @ 17°	HSPF
.	HP 9-411V-4A,5A,6A	CBPH10-490-1P	36,800	7.05	37,600	2.55	22,000	1.90	6.30
.	HP 9-461V-2A	CBPH10-645-2P	39,500	6.85	42,500	2.45	24,800	1.80	6.15
.	HP 9-511V-5A,6A	CBPH10-645-2P	46,500	6.90	53,000	2.65	31,000	1.95	6.65
.	HP 9-511V-5A,6A	CBP10-51-2P	47,000	6.90	53,500	2.70	31,200	1.95	6.70
.	HP 9-651V-2A	CBPH10-1000-1	51,500	6.40	60,500	2.50	37,400	1.95	6.40

Type: HRCU-A-C Trade Name: Fuelmaster +

footnote(s)	outdoor unit	indoor unit	capacity (Btuh)	SEER	capacity (Btuh)	COP @ 47°	capacity (Btuh)	COP @ 17°	HSPF
2	HP 10-211V-1A,2A	C5-495-1FF	19,300	8.60	18,700	2.55	10,100	1.75	6.50
2	HP 10-261V-1A,2A	C5-495-1FF	22,600	8.25	24,200	2.75	14,400	2.05	7.10
2	HP 10-311V-4A,5A	C5-620-1FF	30,000§	7.35§	31,600	2.55§	19,000§	1.85	6.30§
2	HP 10-411V-6A,7A	C5-620-1FF	32,400§	7.95	32,800	2.70	19,200§	2.00	6.90§
.	HP 9-411V-5A,6A	C5-620-1FF	37,000	7.25	37,400	2.60	21,800	1.90	6.35
2	HP 10-511V-6A,7A	C5-805-1FF	46,000§	7.65§	48,500§	2.75§	29,000§	2.00§	6.90
2	HP 11-311/511-2A	C5-805-1FF	47,000	8.95	48,000	2.85	28,400§	2.05	6.95
	HP 9-511V-5A,6A	C5-805-1FF	48,000	7.10	53,500	2.75	31,000	2.00	6.85

LENNOX INDUSTRIES, INC. (CONT.)

Type: HRCU-A-C Trade Name: Lennox

	Model		Cooling		Heating			
footnote(s)	outdoor unit	indoor unit	capacity (Btuh)	SEER	capacity (Btuh)	COP @ 47°	capacity (Btuh)	COP @ 17°	HSPF
2	HP 11-411 /651-2A	C5-920-1FF	54,500	8.55	54,500	2.75	31,400	1.95	6.70

Type: HRCU-A-C Trade Name: Legend

footnote(s)	outdoor unit	indoor unit	capacity (Btuh)	SEER	capacity (Btuh)	COP @ 47°	capacity (Btuh)	COP @ 17°	HSPF
.	HP 11-311/511V-2A	CPH10-645-1	43,000§	8.35§	48,000§	2.65	28,800§	1.95§	6.75
.	HP 11-311/511V-2A	CR12-840-1	47,000§	9.00	49,000§	2.75§	29,000	2.00	6.85§
2	HP 11-311/511V-2A	* C12-840-1	47,500§	9.05§	49,000§	2.80§	29,000§	2.00	6.90§
.	HP 11-311/511V-2A	CP10-51-1	48,000§	9.15§	49,000§	2.75§	29,000	2.00	6.85§
2	HP 11-411/651V-2A	CH10-1000-1	49,500§	8.30§	56,500§	2.40§	34,800§	1.95	6.70§
2	HP 11-411/651V-2A	* C12-1120-1	55,000§	8.85§	58,500§	2.70§	35,600§	2.05§	7.00§
2	HP 11-411/651V-2A	CR12-1120-1	55,500§	8.75§	58,000§	2.65	35,400§	2.00	6.95§

Products not Covered by DOE

Type: HSP-A Trade Name: Lennox

footnote(s)	outdoor unit	indoor unit	capacity (Btuh)	EER	capacity (Btuh)	COP @ 47°	capacity (Btuh)	COP @ 17°	HSPF
	CHP 10-413-2Y	...	33,000	7.50	36,000	2.65	20,000	1.80	...

LENNOX INDUSTRIES, INC. (CONT.)

Type: HSP-A Trade Name: Lennox

	Model		Cooling		Heating				
footnote(s)	outdoor unit	indoor unit	capacity (Btuh)	EER	capacity (Btuh)	COP @ 47°	capacity (Btuh)	COP @ 17°	HSPF
.	CHP 9-413-4Y	...	34,200	6.25	38,500	2.35	22,200	1.60	...
1	CHP 10-463-1G,1Y	...	40,000	7.75	42,500	2.75	23,200	1.85	...
.	CHP 9-463-3Y	...	41,000	6.55	42,500	2.30	26,800	1.65	...
1	CHP 10-513-1G,1Y	...	46,000	7.50	47,000	2.55	27,200	1.75	...
.	CHP 9-513-3Y	...	46,000	6.25	50,000	2.30	30,000	1.60	...
.	CHP 10-653-1G-1Y		54,000	6.55	60,000	2.50	36,400	1.80	...
1	CHP 10B-653-1G,1Y	...	55,000	7.75	57,000	2.70	32,400	1.85	...
.	CHP 11-953-1G,1Q	...	89,000	7.90	89,000	2.80	43,000	1.80	...
.	CHP 11-953-3G,3Q	...	89,000	7.90	89,000	2.80	43,000	1.80	...
.	CHP 11-953-2W	...	89,000	7.80	88,000	2.80	42,000	1.80	...
.	CHP 11-952-3W	...	89,000	7.80	88,000	2.80	42,000	1.80	...
.	CHP 11-1353-3G,3Q	...	122,000	8.05	125,000	2.90	64,000	1.80	...
.	CHP 11-1353-1G,1Q	...	122,000	8.00	125,000	2.90	64,000	1.80	...
.	CHP 11-1353-3W	...	122,000	8.00	123,000	2.90	63,000	1.80	...
.	CHP 11-1353-1W	...	122,000	7.95	123,000	2.90	63,000	1.80	...

LENNOX INDUSTRIES, INC. (CONT.)

Type: HRCU-A-C Trade Name: Lennox

	Model			Cooling			Heating		
footnote(s)	outdoor unit	indoor unit	capacity (Btuh)	EER	capacity (Btuh)	COP @ 47°	capacity (Btuh)	COP @ 17°	HSPF
.	HP 10-413V-2Q,3Q	CPH10-490-1	30,000	7.80	32,400	2.70	18,500	2.00	...
.	HP 10-413V-2Q,3Q	CP12-630-1	32,000	8.00	33,400	2.75	19,500	2.05	...
2	HP 10-413V-2Q,3Q	CR12-630N-1	32,000	8.00	33,400	2.75	19,500	2.05	...
.	HP 10-413V-2Q,3Q	C12-630-1	32,400	8.00	33,400	2.75	19,500	2.05	...
2	HP 10-413V-2Q,3Q	CR12-630-1	32,400	8.00	33,400	2.75	19,500	2.05	...
.	HP 10-413V-2Q,3Q	CP10-41-1	33,400	8.15	34,000	2.80	19,500	2.00	...
.	HP 10-513V-2Q,3Q	CPH10-645-1	43,000	7.55	51,000	2.80	28,400	1.95	...
2	HP 10-513V-2Q,3Q	CR12-840-1	45,000	7.50	50,000	2.80	28,000	1.95	...
.	HP 16-513V-1Y,3Y	CP12-51V-1	46,000	8.55	49,500	3.00	29,600	2.20	...
2	HP 10-513V-2Q,3Q	C12-840-1	46,000	7.65	50,000	2.80	28,000	1.95	...
.	HP 10-513V-2Q,3Q	CP10-51-1	47,000	7.70	50,000	2.80	27,400	1.90	...
.	HP 16-653V-1Y,3Y	CP12-65V-1	56,000	8.40	61,000	3.00	36,400	2.30	...

Type: HRCU-A-CB Trade Name: Lennox

footnote(s)	outdoor unit	indoor unit	capacity (Btuh)	EER	capacity (Btuh)	COP @ 47°	capacity (Btuh)	COP @ 17°	HSPF
	HP 6-953-6Y	CB3-95V	90,000	7.90	93,000	2.80	52,000	1.90	...

LENNOX INDUSTRIES, INC. (CONT.)

Type: HRCU-A-C Trade Name: Fuelmaster +

	Model			Cooling			Heating		
footnote(s)	outdoor unit	indoor unit	capacity (Btuh)	EER	capacity (Btuh)	COP @ 47°	capacity (Btuh)	COP @ 17°	HSPF
2	HP 10-413V-2Q,3Q	C5-620-1FF	32,000	7.90	33,000	2.70	19,000	2.00	...
2	HP 10-513V-2Q,3Q	C5-805-1FF	45,000	7.50	50,000	2.80	27,400	1.90	...

Type: HRCU-A-C Trade Name: Legend

footnote(s)	outdoor unit	indoor unit	capacity (Btuh)	EER	capacity (Btuh)	COP @ 47°	capacity (Btuh)	COP @ 17°	HSPF
2	HP 11-413/653V-2Q	CH10-1000-1	50,000	6.95	55,000	2.50	33,400	1.90	...
2	HP 11-413/653V-2Q	CR12-1120-1	55,000	7.25	57,000	2.65	34,000	2.00	...
2	HP 11-413/653V-2Q	C12-1120-1	56,000	7.30	58,000	2.70	34,400	2.00	...

YORK HEATING & AIR CONDITIONING
DOE Covered Products

Type: HRCU-A-CB Trade Name: Champion

	Model		Cooling			Heating			
footnote(s)	outdoor unit	indoor unit	capacity (Btuh)	SEER	capacity (Btuh)	COP @ 47°	capacity (Btuh)	COP @ 17°	HSPF
.	E2CPO18A06A	N1AHDO6A06/G1AC018AA	18,800	8.30	19,600	2.90	11,200	2.00	7.25
.	E2CP018A06A	* F1CPO18AO6A	19,200	8.30	19,800	3.05	10,800	2.05	6.95
.	E1CPO24AO6A	E1CPO24AO6A	24,000	7.90	26,000	2.90	13,700	1.90	6.90
.	E1CPO24AO6A	* F1CPO24AO6A	25,600	8.10	25,200	2.90	14,000	2.00	6.75
1	E1CPO30AO6A	E1CPO30AO6A	28,000	8.00	30,600	2.80	15,800	1.80	6.85
1	E1CPO30AO6A	* F1CPO30AO6A	28,800	8.00	30,000	2.90	16,800	1.90	6.85
.	E2CPO36AO6A	N2AHD14AO6/G1ACO36AA	34,200	8.60	35,600	3.00	19,000	2.10	7.38
.	E2CPO36AO6A	* F1CPO36AO6A	34,600	8.35	36,400	3.15	20,000	2.15	7.30
.	E1CPO42AO6A	N2AHD14AO6/G1ACO42AA	40,000	8.00	44,000	2.90	25,400	2.10	7.35
1	E2CPO42AO6A	* F1CPO42AO6A	40,500	7.85	43,500	2.95	23,800	2.00	6.90
1	E2CPO48AO6A	* F1CPO48AO6A	45,500	7.60	49,500	2.90	29,000	2.05	6.90
.	E2CPO48AO6A	N2AHD16AO6/G1ACO48AA	46,000	8.00	48,000	2.90	27,200	2.00	7.15
.	E2CPO60AO6A	N2AHD20AO6/G1ACO60AA	59,000	8.30	62,500	2.90	34,400	2.00	7.25
1	E2CPO60AO6A	* F1CPO60AO6A	59,500	7.80	65,000	3.00	37,000	2.15	7.05

YORK HEATING & AIR CONDITIONING (CONT.)

Type: HSP-A Trade Name: Sunpath

	Model		Cooling		Heating				
footnote(s)	outdoor unit	indoor unit	capacity (Btuh)	SEER	capacity (Btuh)	COP @ 47°	capacity (Btuh)	COP @ 17°	HSPF
.	B1SPO24AO6A	...	23,400	8.00	24,600	2.80	13,500	1.90	6.55
.	B1SPO30AO6A	...	28,000	7.50	28,400	2.75	15,200	1.75	6.20
.	B2SPO36AO6A	...	35,000	8.35	34,800	2.90	19,000	1.95	6.85
.	B1SPO42AO6A	...	.38,500	7.25	43,500	2.80	24,000	1.95	6.80
.	B1SPO48AO6A	...	45,000	7.40	47,500	2.70	26,800	1.85	6.50
.	B1SPO60AO6A	...	60,000	7.15	64,000	2.70	37,000	1.90	6.60

Type: HRCU-A-CB Trade Name: Enmod

footnote(s)	outdoor unit	indoor unit	capacity (Btuh)	SEER	capacity (Btuh)	COP @ 47°	capacity (Btuh)	COP @ 17°	HSPF
2,5	E1EMO36A10A	* F1EMO36EO2O1OB	34,800	10.80	37,400	3.00	26,200	1.95	7.80

YORK HEATING & AIR CONDITIONING (CONT.)

Type: HRCU-A-C Trade Name: Champion

	Model		Cooling		Heating				
footnote(s)	outdoor unit	indoor unit	capacity (Btuh)	EER	capacity (Btuh)	COP @ 47°	capacity (Btuh)	COP @ 17°	HSPF
1	E2CPO18AO6A	G1USO18AA	18,800	8.20	19,600	2.90	11,200	2.00	...
.	E2CPO18AO6A	G1CPO18AA	19,000	8.45	19,800	3.00	10,800	2.00	...
1	E1CPO24AO6A	G1USO24AA	24,000	7.80	26,000	2.90	13,700	1.90	...
.	E1CPO24AO6A	G1CPO24AA	25,400	8.05	25,400	2.85	14,200	1.95	...
.	E1CPO30AO6A	G1UNO30AA	28,000	7.70	30,600	2.80	15,800	1.80	...
1	E1CPO30AO6A	G1CPO30AA	28,800	7.85	30,000	2.90	16,800	1.90	...
.	E2CPO36AO6A	G1UNO36AA	34,200	8.40	35,600	3.00	19,000	2.10	...
.	E2CPO36AO6A	G1CPO36AA	34,600	8.40	36,400	3.15	20,000	2.15	...
.	E2CPO42A25A	G1UNO42AA	40,000	8.00	44,000	3.00	25,400	2.10	...
.	E2CPO42AO6A	G1UNO42AA	40,000	7.80	44,000	2.90	25,400	2.10	...
1	E2CPO42A25A	G1CPO42AA	41,000	8.30	43,000	3.10	23,400	2.05	...
1	E2CPO42AO6A	G1CPO42AA	41,000	8.05	43,000	3.00	23,400	2.00	...
.	E2 CPO48A25A	G1UNO48AA	46,000	8.20	48,000	3.00	27,200	2.00	...
.	E2CPO48AO6A	G1UNO48AA	46,000	8.00	48,000	2.90	27,200	2.00	...
1	E2CPO48A25A	G1CPO48AA	46,500	8.45	48,500	3.10	28,000	2.20	...

YORK HEATING & AIR CONDITIONING (CONT.)

Type: HRCU-A-C Trade Name: Champion

	Model		Cooling		Heating				
footnote(s)	outdoor unit	indoor unit	capacity (Btuh)	EER	capacity (Btuh)	COP @ 47°	capacity (Btuh)	COP @ 17°	HSPF
1	E2CPO48AO6A	G1CPO48AA	46,500	8.20	48,500	3.00	28,000	2.15	...
.	E2CPO60A25A	G1UNO60AA	59,000	8.00	62,500	3.00	34,400	2.00	...
.	E2CPO60AO6A	G1UNO60AA	59,000	7.80	62,500	2.90	34,400	2.00	...
1	E2CPO60AO6A	G1CPO60AA	60,000	8.15§	64,500	3.10§	36,400	2.20§	...
1	E2CPO60A25A	G1CPO60AA	60,000	8.15	64,500	3.10	36,400	2.20	...

Type: HRCU-A-CB Trade Name: Champion

footnote(s)	outdoor unit	indoor unit	capacity (Btuh)	EER	capacity (Btuh)	COP @ 47°	capacity (Btuh)	COP @ 17°	HSPF
.	E2CPO42A25A	N2AHD14AO6/G1ACO42AA	40,000	7.90	44,000	2.90	25,400	2.10	...
1	E2CPO42A25A	F1CPO42AOOO6A	41,500	8.20	43,500	3.00	23,800	2.00	...
1	E2CPO48A25A	F1CPO48AO6A	45,500	7.80	49,500	3.00	29,000	2.10	...
.	E2CPO48A25A	N2AHD16AO6/G1ACO48AA	46,000	8.10	48,000	3.00	27,200	2.00	...
.	E2CPO60A25A	N2AHD20AO6/G1ACO60AA	59,000	7.70	62,500	2.80	34,400	1.90	...
1	E2CPO60A 5A	F1CPO60AO6A	59,500	7.90	65,000	3.10	37,000	2.20	...
1	E1CHO90A25A,A46A	F1EHO90A25A,A46A	93,000	8.50	103,000	3.20	59,000	2.50	...
1	E1CH120A25A,A46A	F1EH120A33A	118,000	8.00	129,000	3.00	76,000	2.40	...

YORK HEATING & AIR CONDITIONING (CONT.)

Type: HSP-A Trade Name: Sunpath

	Model		Cooling		Heating				
footnote(s)	outdoor unit	indoor unit	capacity (Btuh)	EER	capacity (Btuh)	COP @ 47°	capacity (Btuh)	COP @ 17°	HSPF
1	B2SPO36A25A	...	35,000	8.10	34,800	3.00	19,000	2.00	...
1	B1SPO42A25A	...	38,500	7.35	43,500	2.90	24,000	2.00	...
1	B1SPO48A25A	...	45,000	7.60	47,500	2.75	26,800	1.90	...
1	B1SPO60A25A	...	60,000	7.55	64,000	2.80	37,000	1.95	...
2	B1SPO90A17B	...	93,000	8.50	101,000	3.10	62,000	2.30	...
1 ,2	B1SPO90A25A	...	95,000	8.50	103,000	3.10	63,000	2.30	...
2	B1SPO90A28B,A46B	...	95,000	8.30	103,000	3.00	63,000	2.40	...
2	B1SP120A17B	...	119,000	8.30	124,000	2.90	75,000	2.30	...
1,2	B1SP120A25B	...	121,000	8.30	126,000	2.80	76,000	2.30	...
2	B1SP120A28B,A46B	...	121,000	8.20	126,000	2.80	76,000	2.30	...

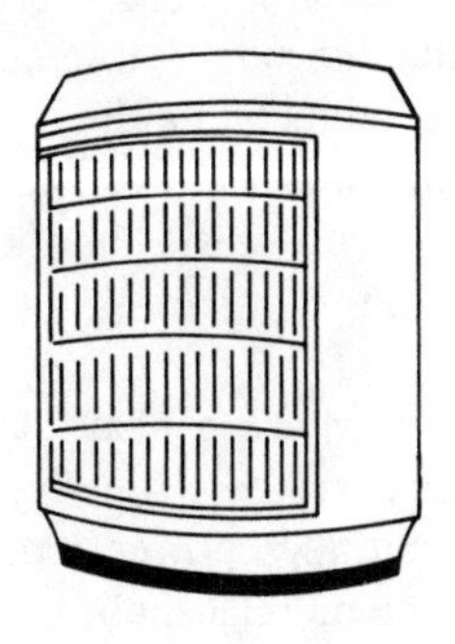

Glossary

A belt—A fan belt used where a heavy load-carrying ability is required. The A belts have the strength to carry a load through short center, small pulley drives. Normally in furnace work the A series replaces the 4L series if a heavy-duty belt is needed.

air filter—A porous article or mass (such as fiberglass or paper) through which air is passed to separate matter in suspension in that air.

alignment—To position two or more parts properly, either mechanical or electrical, in relation to each other. Alignment is used mechanically to reduce friction and wear.

ammeter—An instrument for measuring either direct or alternating current, depending on its internal construction. Its scale is usually graduated in microamperes, milliamperes, amperes, or kiloamperes.

ampere—A unit of electrical current or rate of electron flow. One volt across 1 ohm of resistance causes a current flow of 1 ampere. A flow of 1 coulomb per second equals one ampere. An unvarying current is passed through a solution of silver nitrate of standard concentration at a fixed temperature. A current that deposits silver at the rate of 0.001118 gram per second is equal to 1 ampere, or 6.25 × 10 to the eighteenth power electrons per second is equal to 1 ampere through a circuit.

anticipator—A function of false heating of a thermostatic element to provide more even temperature control in a heated or air conditioned space. Anticipators consist of variable or fixed resistors located within the thermostat assembly. Normally the heating anticipator is adjustable, while the cooling anticipator is of a fixed value. An anticipator is always set for the amount of current passing through the circuit of the thermostat function that the anticipator is connected to in series.

B belt—A fan belt used where a heavy load-carrying ability is required. The B belts have the strength to carry a load through pulleys driven by multihorsepower motors. These belts are usually found only in commercial applications. The A or 4L series belts will not interchange with the B series belts.

backup heat—If the heat pump isn't able to supply all the heat that's required to heat a space in all normal temperature conditions of outside air, backup heat is required. Most installations in colder climates use the furnace to aid the heat pump. See balance point.

balance point—The balance point for a home is computed on an individual basis. Below the balance point temperature, the heat pump will require assistance from the backup heat source to maintain room temperature. Above the balance point temperature, the heat pump alone will maintain room temperature. The balance point temperature will normally fall between 28° F. and 35° F. If the balance point temperature is computed incorrectly, there will be reduced savings, plus possible malfunctioning of the cooling function, because of incorrect sizing of the heat pump.

belt-drive blower—A fan driven by a motor, with a fan belt interconnecting the motor and fan assembly to transfer power to the fan. The belt-driven blower may be called a forced air, belt-driven fan, or simply a blower. The motor and fan assembly contain one pulley each. The pulleys determine fan speed.

belt pulley alignment—The alignment of pulleys to allow a fan belt to run in the plane of the pulley belt grooves, with as little friction as possible. Misaligned pulleys will cause belt noise and premature belt failure. Belt dressing is not advised as a cure; only proper alignment will suffice.

blower housing—The sheet metal enclosure that surrounds the blower wheel to support, align, and direct the flow of air from the blower or forced air fan.

Btu—British thermal unit. A measurement of the quantity of heat from a given source. It is the quantity of heat required to raise the temperature of 1 pound of water 1° F. at or near 39.2° F.

Btuh—British thermal units per hour. Most ratings are given in the per hour number of Btu. Calculations will use the per hour accumulation of Btus as their basis unless stated otherwise.

cad cell—Cadmium sulfide cell. A photoconductive cell in which a small wafer of cadmium sulfide is used to provide an extremely high dark to light resistance ratio. A light sensor used in sensing the presence of fire in oil furnaces.

capacitor—A device consisting essentially of two conducting surfaces separated by an insulating material or dielectric such as air, paper, mica, glass, plastic film, or oil. A capacitor stores electrical energy, blocks the flow of direct current, and permits the flow of alternating current to a degree dependent upon the capacitance and the frequency.

CFM—Cubic feet per minute. A measurement of air flow in cubic feet per minute past a reference point. All properly designed heating systems rely on a specific amount of air flow through each branch of the system. Fans are rated in CFM output under different static pressures or restrictions to air flow.

charge—A term used in refrigeration work, meaning the refrigerant contained within the refrigerant piping system. Charge is measured in pounds of refrigerant.

code—A system of principles or rules; a systematic statement of a body of law. A code is a statement of acceptable mechanical or electrical installation techniques by the local, state, or national governing agencies.

coil—A term used in the trade to identify a heat exchanger. A coil consists of tubing attached to aluminum fins that exchange heat with air. A refrigeration coil looks similar to an automotive radiator.

cold air return—The duct or area that the untreated air passes through to enter the heating or air conditioning equipment. Normally the cold air return would consist of a ductwork system, but not always.

compressor—A machine that compresses gases into a smaller area than the normal state of the gas or free area under atmospheric pressure.

compressor internal high-temperature switch—A temerature-sensitive switch located within the compressor housing as an overload safety device. The internal sensed temperature is at or within the compressor motor where a mechanical or electrical overload would first appear. If overheated, the switch will open the compressor control circuit, cutting the electrical power to the compressor.

compressor internal pressure bypass—Modern compressors incorporate an internal, mechanical safety valve that releases excessive head pressure when the pressure reaches a preset level. The valve is a safety device that protects against possibly dangerous pressure levels.

compressor locked up—A term used to describe a compressor that is locked up internally so that it won't rotate when power is applied to the compressor motor. The compressor will draw excessive current because of its mechanically stalled condition. The compressor will hum noticeably without the characteristic vibration of a running compressor.

compressor valves—Compressors used in air conditioning contain valves to control the flow of gas into and out of the compression chamber. If a valve leaks or otherwise becomes inoperative, the capacity of the compressors will diminish or cease completely.

condensate—Condensed water from the air in a duct system. When air in the duct system passes over a cool surface that's below the dew point temperature, moisture then condenses from that air. The condensed water that comes from the indoor coil is called condensate.

contactor—A heavy-duty relay with one to three poles that is used to switch large amounts of electrical power. Magnetic contactors are used in most air conditioning systems to turn the compressor and condensing fan on and off. A pole is one switching circuit; a light switch in a home is a single-pole device.

contacts—Relays, contactors, light switches, or any devices used for turning electrical power on and off contain a mechanical point of contact that completes the electrical circuit. This point of contact is coated with a special antipitting metal that retards corrosion on the contact surface. When the special metal is worn away due to electrical arcing, the contacts will fail to make a circuit or weld together.

cooling anticipator—A function of false heating of a thermostatic element to provide more even temperature control in a heated or air conditioned space. Normally the cooling anticipator is nonadjustable. The anticipator resistor generates a small amount of heat to turn the thermostat on slightly sooner than if that thermostat was only sensing room heat. The anticipator reduces the lag time of the cooling system.

damper—A dulling or deadening device. In the heating and air conditioning trade, a damper is used inside the air distribution pipework to retard the air flow from the system. Dampers have many forms—from a flat plate to multifinned register dampers.

defective—Used to describe a part or system that has become inoperative as to its designed function.

defrost cycle—A mode of operation common to heat pumps that melts frost and ice from the outdoor section of a heat pump—normally automatically. The defrost cycle is not needed in units that use water as a heat source. A defrost cycle in a heat pump is cooling without the operation of the outdoor fan. The defrost cycle is started by a time-temperature base or by electronic means. The cycle is terminated by temperature in the outdoor section of the heat pump or by a built-in time limit.

defrost relay—An electrical switch used to make wiring changes required to perform the defrost function on command within the heat pump circuitry.

defrost thermostat—Temperature sensor used to detect the temperature of the outdoor unit coil. When the temperature of that coil reaches a predetermined level, the defrost cycle will start. The heat pumps that use a time temperature defrost control system terminate the defrost cycle after the outdoor coil has reached approximately 75° to 80° F., or if about 10 minutes have passed with the machine in defrost. Electronic control of the defrost cycle is started by measuring the outdoor temperature and the temperature of the outdoor coil. The electronic termination of defrost is also on a time temperature base. The electronic type of defrost circuit is considered more efficient by many technicians and manufacturers.

defrost timer—An electrical mechanical device or electronic circuit used to determine the defrost cycle if the outdoor coil isn't cleared of ice or frost within a predetermined time limit. If the defrost thermostat should fail, the defrost cycle always will last for the full timed duration of the heat pump timer, thus lowering efficiency.

device—A piece of equipment or a mechanism designed to serve a special purpose or perform a special function.

diffuser—A device for reducing the velocity and increasing the static pressure of a fluid passing through a system. A device (as slats at different angles) for deflecting air from an outlet in various directions.

direct-drive blower—A furnace forced air fan assembly that is electrically driven by a motor connected directly to the motor shaft. Because no pulley arrangement is used, the RPM rating of the fan is dependent on the power and RPM rating of the motor.

downflow furnace—A furnace that discharges the conditioned air directly under the main body of the furnace.

electronic air filter—A special type of air filter that is highly efficient when compared to other types of filter mediums. The electronic filter has an efficiency of 90 percent or better, consumes little power, and is washable.

electronic defrost sensor—A sensor known as a thermistor that detects and reacts electrically to changes in temperature. Few heating and air conditioning manufacturers use the electronic sensor because of the higher than mechanical cost.

element—The wire that furnishes heat within an electric furnace. Elements in electric furnaces are normally rated in kilowatts. A 5-kW element is the norm at 240 volts supply. Most furnaces contain more than two elements. A 20-kW furnace will have 4 elements.

end play—The amount that a shaft (such as a horizontal fan shaft) travels from side to side in a bearing surface before being mechanically stopped from moving further.

fan—A device for producing a current of air that consists of a series of vanes rotated on an axle driven by a motor.

fan assembly—The fan, fan support, shroud, and motor. The fan assembly is a complete air-moving device and is a subsystem of a furnace or air handler.

fan belt tension—The degree of tightness observed or measured when a fan belt is mounted on its pulley arrangement. Proper belt tension is important when belt life is considered.

fan switch—An electrical switch that supplies electrical power to a fan. Fan switches operate either automatically or manually. Automatic fan switches find their application in turning the furnace fan on and off at specific, manually set temperatures. Manual fan switches are used to permit fan operation continuously by placing a switch in the on position. A fan switch is usually single-pole, single-throw.

filter contamination—The effect of a filter collecting any material being carried by air as a medium of transfer. Contamination materials are solid rather than gaseous. Dust, lint, and smoke are examples of everyday contamination.

fin—A small piece of metal attached to the tubes in a coil for the purpose of increasing the heat transfer area of that tube. Each coil will have hundreds of these aluminum fins to greatly increase the radiation of the coil or its ability to transfer temperature.

fin comb—A tool fabricated from nylon or plastic for the purpose of straightening bent fins in an air conditioning coil. The tool will have a series of slots cut into a bar of nylon or plastic that match the fin density of a particular coil. Five or six fin densities are presently in general use, and the density is measured in the number of fins per inch.

flexible joint—A joint used in sheet metal work that provides mechanical isolation between two sections of duct material. A flexible joint consists of a piece of fabric (sometimes rubber-covered fabric) with metal edges to connect the joint to the ductwork. A flexible joint will give from ¼″ to 2″ of mechanical separation.

four-way valve—An electromechanical device used to reverse the flow of refrigerant from the heating to cooling mode of operation. The four-way valve (reversing valve) is switched electrically and has four pipe connections.

fractional horsepower belt—A fan belt rated for light-duty applications such as furnace or air handler blowers. The 4L series belts are rated as fractional horsepower in nature. Motors from 1/6 horsepower to 1/2 horsepower normally use fractional horsepower belts. Excessive belt strength causes excessive friction or power to roll the belt over the pulleys.

fractional horsepower motor—An electric motor rated at less than 1 horsepower of mechanical ability.

free travel—The amount a device or part is allowed to move between adjustments, stops, thrust collars, or in an untensed mode of movement.

fusible link—A temperature device used in electric and gas furnaces to stop the unit from operating if excessive temperature is detected. The fusible link is nothing more than a metal with a low melting point inserted into an electrical circuit that melts away when its temperature limit is met, thus opening the electrical connection.

fusible plug—A small piping plug with a hole drilled in its center. The hole is filled with a metal that has a low melting point. The plug is mechanically fit to a possible problem area that may overheat. If the object in question overheats, the metal liquifies, thus opening the hole in the center of the plug. The fusible plug is used to prevent excessive temperature and pressure conditions within a heat pump or air conditioner refrigerant circuit.

gas—A fluid (as air) that has neither independent shape nor volume but tends to expand indefinitely.

gas, natural—A combustible mixture of methane and higher hydrocarbons used chiefly as a fuel and raw material. One cubic foot of natural gas contains approximately 1,000 to 1,100 Btu. It is normally used in gas furnaces.

gas valve—A valve that controls natural gas flow into a heating device. Depending on the type of valve, it may or may not include a safety shutoff for the gas.

gas valve, main—A main gas valve will include an operator (a device to turn the gas flow on by command), a thermocouple type safety that overrides a command to turn on if the pilot has malfunctioned, and the main body of the gas valve. The main body contains the primary valve to turn the gas on and off and the passages necessary for operation. The valve also normally includes a manual shutoff valve that can be set in the off position, pilot operation only, and on for complete functioning of the valve. Thermocouple voltages for the safety circuit are varied, depending on the application. Operator voltages are also varied, depending on the application. The voltages are as follows:

Thermocouple safety voltages	*Operator control voltages*
0.030 volts, dc	0.5 to 0.75 volts, dc

Thermocouple safety voltages	*Operator control voltages*
0.500 volts, dc	24 volts, ac
0.750 volts, dc	120 volts, ac

Special types of thermostats are required for the different operator control voltages.

head pressure—The pressure that a refrigeration compressor delivers from the discharge of that compressor before the refrigerant is condensed into a liquid. The condenser liquid pressure is called the high-side pressure, with high side meaning the higher of compressor pressures.

heat anticipator—A function of false heating of a thermostatic element to provide more even temperature control in a heated or air conditioned space. Normally the heating anticipator is adjustable by hand. The anticipator resistor generates a small amount of heat to turn off slightly sooner than if the thermostat was only sensing room temperature. The anticipator reduces the overshoot experienced by nonanticipated systems. The setting of a heating anticipator should match the current flowing through the thermostat.

heat gain—The amount of heat gained from outside sources such as the sun, people, machinery, and lighting in a living space. The heat gain figure must be known to compute the size of a cooling unit for that space to the maximum design temperature.

heating plant—The source of heating in a living space is sometimes called a heating plant, furnace, stove, or stoker.

heat loss—The amount of heat lost from a living space through the walls, windows, floor, and roof. The amount of heat lost from a space must be known to size a heating plant to the minimum design temperature.

heat pump—An apparatus for heating or cooling a building by transferring heat by mechanical means from or to a reservoir (as the ground, water, or air) outside the building.

high side—The term used to describe the pressure output of a refrigeration compressor that is flowing away from the compressor. The high-side pressure is the opposite of the low-side pressure.

high-side restriction—A blockage of the liquid line curtailing the flow of refrigerant in the system. Restrictions may be found by feeling the temperature of the liquid line. If the blockage is severe enough, the line will feel warm, then suddenly turn cold at the point of the blockage.

horizontal furnace—A furnace that is longer than the total height of the unit. The incoming air enters on end, travels through the furnace, and is discharged from the other end.

horsepower—A unit of power or the capacity of a mechanism to do work. It is the equivalent of raising 33,000 lbs. 1 foot in one minute or 550 pounds 1 foot in one second. One horsepower equals 746 watts of electrical power.

humidity—The amount of water in a given volume of air at a specific temperature.

humidity sensor—A sensor designed to measure the amount of humidity in the air surrounding the sensor. The sensing element used commonly is a special type of plastic that changes shape with varying degrees of wetness.

icing—A term used to describe the effect of ice on a refrigeration device. Cooling or air conditioning parts shouldn't ice, but a freezer will if it functions properly.

indoor fan—The furnace or air handler fan that delivers air to the interior of the building.

instantaneous current—The magnitude of electrical current, at any particular instant, from a varying value. Instantaneous current values may be thousands of times higher than the operating current of a device.

instantaneous voltage—The magnitude of electrical voltage, at any particular instant, from a varying value. Instantaneous voltage values may be thousands of times higher than the operating voltage of a device. The collapse of a magnetic field with its induced voltage effect is an example.

insulation—A material such as fiberglass, rock wool, treated paper, or foam that impedes the flow of heat through that material. All materials have an insulating value, but some of these materials are much better than others.

limit—A safety control that protects against excessive internal temperature within the furnace. Limit switches opening the electrical circuit to the fire control.

load—The power consumed by a machine or circuit in performing its function. A load can be mechanical or electrical in nature. An example of a mechanical load is the heat in a room that an air conditioner is removing. An electrical load is the electrical power that the air conditioner consumes or the amount of power used by the machine.

lowboy furnace—A type of furnace that has the air intake (cold air return) and the air output (the supply air) on the same elevation on the furnace casing. It is an older design that is not used commonly in today's heating units.

low side—A term used in refrigeration work that describes the lesser pressure side of the refrigeration circuit. The low side is the pressure entering the evaporator to and including the pressure sensed at the compressor intake valve or valves.

low voltage—In heating and refrigeration work, the term usually refers to a 24-volt circuit. Twelve or 20-volt circuits are applied in special cases. Alternating current (ac) is used with 24-volt circuits. The 12- and 20-volt circuits are generally direct current in nature (dc).

low-voltage fuse—A fuse used in an electrical circuit in a heating or air conditioning application that, protects a low-voltage current path in that electrical circuit.

metering device—A device that controls the flow of refrigerant in the refrigerant circuit. The device is a capillary tube or a thermostatic expansion valve. The capillary tube is a small copper tube with a small-diameter hole in its center that will allow only so much refrigerant to flow under a certain pressure because of the restriction the hole presents to the flowing liquid. The expansion valve senses the temperature at a specific point in the refrigeration circuit and attempts to maintain that temperature by increasing or decreasing the refrigerant flow automatically. It is a temperature-reactive device.

mode—The type of operation that a machine is responding to for a specific function. Heating, cooling, and defrost are the modes of operation for a heat pump.

motor oil—An oil designed for automotive use. It is used as a lubricant for electric motors and forced air fans. Nondetergent 20-weight motor oil is recommended for oiling motors.

multiple sequencer—A sequencer with multiple sections designed to handle more than one electric heating element at one time within an electric furnace. Several physical differences exist among manufacturer's designs, but the functions are the same. More than one electrical switch in an assembly is used to control electric furnace heating elements.

multiple stage—A trade term used to describe a heating or air conditioning unit that has more than one level of output to the conditioned space.

nozzle, oil—A device used in an oil furnace that sprays oil under pressure (100 psi or more in a high-pressure burner) in a specific spray pattern. Nozzles are rated in gallons of oil per hour at 100 psi pressure. A subrating is the spread of the spray from the tip of the nozzle in degrees, ranging from 30° to a 90° spray angle. The center of the cone of sprayed oil is also rated as hollow, semisolid, or solid, meaning that the inner section of the cone sprayed has a degree of droplet density when compared to the shell of the cone. An example is a nozzle with an outer cone that represents a 100 percent density, but with not as heavy a concentration in the center of the cone. This is a semisolid nozzle. Different burners require different nozzles for their highest efficiency level.

ohmmeter—A direct reading instrument for measuring electric resistance. Its scale is usually graduated in ohms, kilohms, or both. It is the preferred instrument to check fuses for a current path.

oil, fuel—Fuel oil for residential furnaces comes in number one and number two grades. Number one is commonly referred to as stove oil. Number two is known as furnace oil or diesel fuel. The Btu content of furnace oil is generally accepted as being 138,000 Btus per gallon.

oil pump—A pump used in oil heating systems that removes oil from the oil tank, increases its pressure, and regulates the pressure being delivered to the oil spray system.

oil pump coupling—A device used to interconnect the oil pump and the oil pump motor, usually manufactured from rubber, with metallic end caps. It provides a mechanical connection and compensates for minor misalignment of the motor and pump.

open—An electrical term meaning that current can not flow because the circuit or wire is broken. An open circuit is open by design or fault.

outdoor fan—A fan located in the outdoor unit of a heat pump or air conditioner. If it is part of an air conditioner, it could be referred to as a condensing fan. If in a heat pump, the fan has a dual purpose and is only referred to as the outdoor unit fan.

outdoor thermostat—A thermostat used to turn the backup heat off if the temperature outside is above a preset point. More than one outside thermostat is common.

outdoor unit—The outdoor section of a heat pump or air conditioner in a split heating or cooling installation.

overload—A mechanical or electrical term used to describe the overworking of a device beyond its rated capacity.

power—The rate at which work is done. Units of power are the watt, joule, and kilowatt.

power factor—The ratio of the actual power of an alternating or pulsating current, as measured by a wattmeter, to the apparent power as indicated by an ammeter and voltmeter. Electric motors have a power factor (PF) rating.

precharged line set—A set of two copper pipes that interconnect the outdoor unit with the indoor coil. The larger of the two lines is insulated with black foam rubber to stop condensation of water moisture or heat loss. The smaller of the two lines, the liquid line, is not covered. A precharged line set is manufactured and sold as a kit. Refrigerant and the line connectors are included to reduce installation time.

primary control—An oil furnace control that activates the fire from a command from the thermostat. The primary control also contains the fire safety. If the fire fails to light or goes out, the safety will turn the burner off in 30 to 90 seconds. Oil furnace primary controls use heat or light sensing for safety control.

primary control safety—Oil primary control safety switches sense heat or light to determine the presence of fire. The sensors are a cad cell or the helix mechanism driving the safety switch circuit.

qualified serviceman—A person that has attained training and field experience to be effective as a equipment serviceman. Four or more years are usually required for training in this trade.

refrigerant—A chemical that has the ability to provide cooling or heating when it's mechanically processed. Refrigerant can be liquid or a gas, depending on temperature requirements.

refrigerant charge—The amount of refrigerant in a heating or cooling system.

refrigerant flow—The refrigerant gas or liquid flowing through the piping or compressor system.

register—A device that directs the flow of air from an air pipe in a duct system. Registers are floor-mounted, wall-mounted, and ceiling-mounted. They are sometimes called diffusers (they diffuse air flow).

relay—An electromechanical device in which contacts are opened and/or closed by variations in the conditions of one electric circuit, thereby affecting the operation of other devices in the same or other electric circuits. A relay consists of an activation coil and a set(s) of contacts. Normally the coil in a heating/cooling application will be activated by 24 volts or 120 volts ac.

resistance—An electrical property that impedes the flow of current in a circuit.

resistive connection—Any kind of electrical connection that has a layer of oxide or contamination between the contact points. The oxide will inhibit the flow of current, which means that the current to the supplied circuit will cease or be reduced in value. In high-current circuits, the formation of a resistive connection almost always burns the connecting wire, thus opening the circuit. Resistive connections are known to generate heat.

resistor—A device that resists the flow of electric current in a circuit. There are fixed and variable resistors.

return duct—A duct that carries room air back to the air handler or furnace for processing. A duct is nothing more than a confined air passage.

return grille—A device used to cover a wall opening that is connected to the return air duct. A finishing device to cover a crude hole in a wall.

return plenum—A large duct that returns air from the cold air (return air) ducting to the furnace or air handler.

reversing valve—See four-way valve.

RPM—The rate that a mechanical device rotates.

run capacitor—A capacitor that is used with electric motors to lower the amount of running current. The capacitor improves the power factor of the motor.

R value—A rating of insulating value. The higher the number, the better the material insulates.

sequencer—A special control used primarily to control the on and off function of electric heating elements in furnaces or air handlers. The sequencer is usually activated by a 24-volt control signal fed to a thermal warp switch. The warp switch closes contacts that complete a circuit to the heating element. Sequencers usually are supplied with one-, two-, three-, four-, and five-element switches, plus low current auxiliary switch functions. Sequencers display a time delay between the control signal and main circuit activation.

service—The act of testing or repairing a mechanical, electrical, or electromechanical device.

shorted—Also called a short circuit. An abnormal connection of relatively low resistance between two points of a circuit. The result is a flow of excessive (often damaging) current between these points.

sizing—A term used in the industry to describe computing the heating or cooling size of a unit in Btu for a particular application.

slow blow fuse—A special type of fuse capable of carrying a current in excess of its rating for a short period. The fuses protect motors. Motors draw a higher than average current during the start condition. Belt-drive blowers can draw a current 10 times their normal running current on start.

snow base—A mechanical stand or any device to raise the elevation of a heat pump so that snow will not exceed the lower level of the heat pump base.

spec house—A house or home built to the specification of an architect.

split-phase motor—An electrical motor used in heating and air conditioning work to drive belt-drive blowers.

split system—A term used to describe a heat pump or air conditioning system that has its components in two primary locations interconnected by control wires and refrigeration piping.

spreading out of fan shaft material—If a setscrew is tightened on the round portion of a fan shaft, the material under the setscrew will expand, causing the collar containing the setscrew to be mechanically locked in place by expanded shaft material. Setscrews should only be tightened on a flat section of shaft.

stage balance point—A computation used in heat pump design where different levels of outside temperature trigger additional levels of backup heating to compensate for losses from the home.

stop collar—A collar used to prevent sideways motion of a fan shaft to definite mechanical limits.

stuck fan relay—Electrical contacts are welded or mechanically together in a way that the circuit won't open.

supply duct—A duct that supplies heated or air conditioned air to a living space.

supply plenum—Normally a vertical section of duct that connects the furnace or air handler to the supply ductwork in a heating or air handling system. Plenum means full.

thermal link—See fusible link.

thermocouple—Also called a thermal junction. A device for measuring temperature where two electrical conductors of dissimilar metals are joined at the point of heat application. A resulting voltage difference, directly proportional to the temperature, is developed across the free ends and is measured potentiometrically. A safety device used in gas furnaces.

thermostat—An electrical switch that is turned on and off by the sensing of temperature in a specific area.

thrust collar—See stop collar.

ton—A term used in air conditioning and refrigeration work meaning that a device can deliver 12,000 Btuh. Twelve thousand Btuh equal 1 ton of cooling.

transition—A section of ductwork that changes duct size so that it will fit a duct of different sizing or configuration.

turn-on time—The amount of time required to turn a device on after the turn-on command is given by the control circuit.

unobstructed airflow—Air flow through a space that does not create friction to slow the air or to impede its designed travel rate.

upflow furnace—The supply air is discharged from the top of the furnace, and the return air is drawn into the bottom of the furnace.

U value—An insulating value similar to R value. The U value is an older designation seldom used now.

vent—A term used for a part used on a gas furnace. The chimney of a gas furnace or a pipe leading to that chimney. Gas vents are usually made of lighter material than those of oil, coal, or wood appliances.

volt—A difference of electrical potential between two measured points. Voltage is the amount of electrical pressure that determines amp flow between two points in a circuit.

watt—Abbreviated W. A unit of electric power required to do work at the rate of 1 joule per second. It's the power expended when 1 ampere of direct current flows through a resistance of 1 ohm. In an alternating current circuit the true power in watts is effective volts multiplied by the circuit power factor. There are 746 watts in 1 horsepower.

Index

Noted 2-2-93

Edited by Robert Ostrander